METHODS FOR **DEVELOPING SPACECRAFT WATER EXPOSURE** GUIDELINES

Subcommittee on Spacecraft Water Exposure Guidelines
Committee on Toxicology
Board on Environmental Studies and Toxicology
Commission on Life Sciences
National Research Council

NATIONAL ACADEMY PRESS
Washington, D.C.

NATIONAL ACADEMY PRESS2101 Constitution Ave., N.W. Washington, D.C. 20418

NOTICE: The project that is the subject of this report was approved by the Governing Board of the National Research Council, whose members are drawn from the councils of the National Academy of Sciences, the National Academy of Engineering, and the Institute of Medicine. The members of the committee responsible for the report were chosen for their special competences and with regard for appropriate balance.

This project was supported by Grant No. NAG 9-895 between the National Academy of Sciences and the National Aeronautics and Space Administration. Any opinions, findings, conclusions, or recommendations expressed in this publication are those of the author(s) and do not necessarily reflect the view of the organizations or agencies that provided support for this project.

International Standard Book Number 0-309-07134-8

Additional copies of this report are available from:

National Academy Press
2101 Constitution Ave., NW
Box 285
Washington, DC 20055

800-624-6242
202-334-3313 (in the Washington metropolitan area)
http://www.nap.edu

Printed in the United States of America

THE NATIONAL ACADEMIES

National Academy of Sciences
National Academy of Engineering
Institute of Medicine
National Research Council

The **National Academy of Sciences** is a private, nonprofit, self-perpetuating society of distinguished scholars engaged in scientific and engineering research, dedicated to the furtherance of science and technology and to their use for the general welfare. Upon the authority of the charter granted to it by the Congress in 1863, the Academy has a mandate that requires it to advise the federal government on scientific and technical matters. Dr. Bruce M. Alberts is president of the National Academy of Sciences.

The **National Academy of Engineering** was established in 1964, under the charter of the National Academy of Sciences, as a parallel organization of outstanding engineers. It is autonomous in its administration and in the selection of its members, sharing with the National Academy of Sciences the responsibility for advising the federal government. The National Academy of Engineering also sponsors engineering programs aimed at meeting national needs, encourages education and research, and recognizes the superior achievements of engineers. Dr. William A. Wulf is president of the National Academy of Engineering.

The **Institute of Medicine** was established in 1970 by the National Academy of Sciences to secure the services of eminent members of appropriate professions in the examination of policy matters pertaining to the health of the public. The Institute acts under the responsibility given to the National Academy of Sciences by its congressional charter to be an adviser to the federal government and, upon its own initiative, to identify issues of medical care, research, and education. Dr. Kenneth I. Shine is president of the Institute of Medicine.

The **National Research Council** was organized by the National Academy of Sciences in 1916 to associate the broad community of science and technology with the Academy's purposes of furthering knowledge and advising the federal government. Functioning in accordance with general policies determined by the Academy, the Council has become the principal operating agency of both the National Academy of Sciences and the National Academy of Engineering in providing services to the government, the public, and the scientific and engineering communities. The Council is administered jointly by both Academies and the Institute of Medicine. Dr. Bruce M. Alberts and Dr. William A. Wulf are chairman and vice chairman, respectively, of the National Research Council.

SUBCOMMITTEE ON SPACECRAFT WATER EXPOSURE GUIDELINES

DONALD E. GARDNER *(Chair)*, Consultant, Raleigh, North Carolina
JOSEPH V. BRADY, Johns Hopkins University, Baltimore, Maryland
GARY P. CARLSON, Purdue University, West Lafayette, Indiana
ELAINE M. FAUSTMAN, University of Washington, Seattle, Washington
CHARLES E. FEIGLEY, University of South Carolina, Columbia, South Carolina
MARY ESTHER GAULDEN, University of Texas Southwestern Medical Center, Dallas, Texas
WILLIAM E. HALPERIN, National Institute for Occupational Safety and Health, Cincinnati, Ohio
RALPH L. KODELL, National Center for Toxicological Research, Jefferson, Arkansas
ROBERT SNYDER, Environmental and Occupational Health Sciences Institute, Piscataway, New Jersey
BERNARD M. WAGNER, Consultant, Short Hills, New Jersey
GAROLD S. YOST, University of Utah, Salt Lake City, Utah

Staff

SUSAN N.J. PANG, Project Director
LEE R. PAULSON, Project Director (through 10/99)
KATE KELLY, Technical Editor
KATHRINE IVERSON, Information Specialist
LUCY V. FUSCO, Project Assistant
EMILY L. SMAIL, Project Assistant

Sponsor

NATIONAL AERONAUTICS AND SPACE ADMINISTRATION

Committee on Toxicology

Board on Environmental Studies and Toxicology

Commission on Life Sciences

OTHER REPORTS OF THE BOARD ON ENVIRONMENTAL STUDIES AND TOXICOLOGY

Toxicological Effects of Methylmercury (2000)
Strengthening Science at the U.S. Environmental Protection Agency: Research-Management and Peer- Review Practices (2000)
Scientific Frontiers in Developmental Toxicology and Risk Assessment (2000)
Modeling Mobile-Source Emissions (2000)
Toxicological Risks of Selected Flame-Retardant Chemicals (2000)
Copper in Drinking Water (2000)
Ecological Indicators for the Nation (2000)
Waste Incineration and Public Health (1999)
Hormonally Active Agents in the Environment (1999)
Research Priorities for Airborne Particulate Matter: I. Immediate Priorities and a Long-Range Research Portfolio (1998); II. Evaluating Research Progress and Updating the Portfolio (1999)
Ozone-Forming Potential of Reformulated Gasoline (1999)
Risk-Based Waste Classification in California (1999)
Arsenic in Drinking Water (1999)
Brucellosis in the Greater Yellowstone Area (1998)
The National Research Council's Committee on Toxicology: The First 50 Years (1997)
Toxicologic Assessment of the Army's Zinc Cadmium Sulfide Dispersion Tests (1997)
Carcinogens and Anticarcinogens in the Human Diet (1996)
Upstream: Salmon and Society in the Pacific Northwest (1996)
Science and the Endangered Species Act (1995)
Wetlands: Characteristics and Boundaries (1995)
Biologic Markers (5 reports, 1989-1995)
Review of EPA's Environmental Monitoring and Assessment Program (3 reports, 1994-1995)
Science and Judgment in Risk Assessment (1994)
Ranking Hazardous Waste Sites for Remedial Action (1994)
Pesticides in the Diets of Infants and Children (1993)
Issues in Risk Assessment (1993)
Setting Priorities for Land Conservation (1993)
Protecting Visibility in National Parks and Wilderness Areas (1993)
Dolphins and the Tuna Industry (1992)
Hazardous Materials on the Public Lands (1992)
Science and the National Parks (1992)
Animals as Sentinels of Environmental Health Hazards (1991)

Assessment of the U.S. Outer Continental Shelf Environmental Studies Program, Volumes I-IV (1991-1993)
Human Exposure Assessment for Airborne Pollutants (1991)
Monitoring Human Tissues for Toxic Substances (1991)
Rethinking the Ozone Problem in Urban and Regional Air Pollution (1991)
Decline of the Sea Turtles (1990)

Copies of these reports may be ordered from
the National Academy Press
(800) 624-6242
(202) 334-3313
www.nap.edu

OTHER REPORTS OF THE COMMITTEE ON TOXICOLOGY

Review of the U.S. Navy's Exposure Standard for Manufactured Vitreous Fibers (2000)

Submarine Exposure Guidance Levels for Selected Hydrofluorocarbons: HFC-236fa, HFC-23, and HFC-404a (2000)

Review of the U.S. Army's Health Risk Assessments for Oral Exposure to Six Chemical-Warfare Agents (1999)

Toxicity of Military Smokes and Obscurants, Volume 1 (1997), Volume 2 (1999), Volume 3 (1999)

Assessment of Exposure-Response Functions for Rocket-Emission Toxicants (1998)

Review of a Screening Level Risk Assessment for the Naval Air Facility at Atsugi, Japan (Letter Report) (1998)

Toxicity of Alternatives to Chlorofluorocarbons: HFC-134a and HCFC-123 (1996)

Permissible Exposure Levels for Selected Military Fuel Vapors (1996)

Spacecraft Maximum Allowable Concentrations for Selected Airborne Contaminants, Volume 1 (1994), Volume 2 (1996), Volume 3 (1996), Volume 4 (2000)

Preface

THE National Aeronautics and Space Administration (NASA) maintains an active interest in the environmental conditions associated with living and working in spacecraft and identifying hazards that might adversely affect the health and well-being of crew members. Despite major engineering advances in controlling the spacecraft environment, some water and air contamination appears to be inevitable. Several hundred chemical species are likely to be found in the closed environment of the spacecraft, and as the frequency, complexity, and duration of human space flight increase, identifying and understanding significant health hazards will become more complicated and more critical for the success of the missions.

NASA asked the National Research Council (NRC) Committee on Toxicology to develop guidelines, similar to those developed by the NRC in 1992 for airborne substances, for examining the likelihood of adverse effects from water contaminants on the health and performance of spacecraft crews. In this report, the Subcommittee on Spacecraft Water Exposure Guidelines (SWEGs) examines what is known about water contaminants in spacecraft, the adequacy of current risk assessment methods, and the toxicologic issues of greatest concern. SWEGs are to be established for exposures of 1, 10, 100, and 1000 days. The 1-day SWEG is a concentration of a substance in water that is judged to be acceptable for the performance of specific tasks during rare emergency conditions lasting for periods up to 24 hours. The 1-day SWEG is intended to prevent irreversible harm and degradation in crew performance. Temporary discomfort is permissible as long as there is no effect on judgment, performance, or ability to respond to an emergency. Longer-term SWEGs are intended to prevent adverse health effects (either immediate or delayed) and degradation in crew performance that could result from continuous exposure in closed spacecraft for as long as 1000 days.

This report has been reviewed in draft form by individuals chosen for their technical expertise and diverse perspectives, in accordance with procedures approved by the NRC Report Review Committee for reviewing NRC and Institute of Medicine reports. The purposes of that independent review were to provide candid and critical comments to assist the NRC in making the published report as sound as possible and to ensure that the report meets institutional standards for objectivity, evidence, and responsiveness to the study charge. The review comments and draft manuscript remain confidential to protect the integrity of the deliberative process. We wish to thank the following individuals, who are neither officials nor employees of the NRC, for their participation in the review of this report: Joseph Borzelleca, Virginia Commonwealth University; Dean Carter, University of Arizona; John Doull, The University of Kansas Medical Center; Rogene Henderson, Lovelace Respiratory Research Institute; Robert Kavlock, U.S. Environmental Protection Agency; and Robert MacPhail, U.S. Environmental Protection Agency.

The individuals listed above have provided many constructive comments and suggestions. It must be emphasized, however, that responsibility for the final content of this report rests entirely with the authoring subcommittee and the NRC.

Special thanks are extended to Dr. Raghupathy Ramanathan of Wyle Laboratories, who provided the critical background information in Chapter 2. Thanks are also extended to Drs. Hector Garcia, Chiu-Wing Lam (both from Wyle Laboratories), and John T. James (Johnson Space Center), who provided technical information for the report. We gratefully acknowledge the staff at the Water and Food Analysis Laboratory at NASA's Johnson Space Center for their support in providing water analysis data and help with matters concerning *Mir* water systems. In particular, we wish to thank Mr. Dick Sauer, Dr. John Schultz, Dr. Paul Mudgett, and Mr. John Straub.

We are grateful for the assistance of the NRC staff in preparing the report. Staff members who contributed to this effort are Carol Maczka, senior program director for Toxicology and Risk Assessment; Kulbir Bakshi, program director for the Committee on Toxicology; and Kate Kelly, editor. We especially wish to recognize the contributions of project directors Lee Paulson and Susan Pang and project assistants Lucy Fusco and Emily Smail.

Finally, we would like to thank all the members of the subcommittee for their dedicated efforts throughout the development of this report.

Donald E. Gardner, Ph.D.
Chair, Subcommittee on Spacecraft Water Exposure Guidelines

Bailus Walker Jr., Ph.D., M.P.H.
Chair, Committee on Toxicology

Contents

ABBREVIATIONS xix

EXECUTIVE SUMMARY 1

1 INTRODUCTION 9
Water Contaminants, 10
Approach to the Study, 12
References, 12

2 SOURCES, TREATMENT, AND MONITORING OF SPACECRAFT WATER CONTAMINANTS 14
Overview, 14
Sources of Spacecraft Water Contamination, 18
Regenerated Water, 37
Monitoring Water Contaminants, 48
Summary, 51
References, 53

3 SOURCES AND TYPES OF DATA FOR ESTABLISHING SPACECRAFT WATER EXPOSURE GUIDELINES 57
Chemical and Physical Characteristics of a Toxicant, 58
Human Studies, 59

Animal Studies, 60
In Vitro Toxicity Studies, 63
Advances in Health Effects Assessment, 63
Mechanistic Studies, 69
Summary, 70
References, 71

4 RISK ASSESSMENT METHODS FOR DETERMINING SPACECRAFT WATER EXPOSURE GUIDELINES 75
Historical Perspective, 76
Recommended Approach to Risk Assessment, 79
BMD Calculation, 82
Exposure Conversion, 84
Uncertainty Factors, 87
Setting SWEGS, 93
Alternative Approaches, 97
Summary, 100
References, 101

5 RANKING SPACECRAFT CONTAMINANTS FOR RISK ASSESSMENT 108
Approaches to Risk Prioritization, 109
Flexibility of Risk Prioritization, 112
Summary, 112
References, 113

APPENDIX A: WATER RECLAMATION SYSTEMS ON *MIR* AND THE INTERNATIONAL SPACE STATION 115

APPENDIX B: BENCHMARK DOSE ESTIMATION 131

APPENDIX C: BIOGRAPHICAL INFORMATION 149

Abbreviations

ADI	acceptable daily intake
BBDR	biologically based dose-response
BMD	benchmark dose
CFU	colony-forming unit
CHeCS	crew health-care system
CWC	contingency water container (U.S.)
ECLSS	environmental control life support system
EPA	U.S. Environmental Protection Agency
EVA	extravehicular activity
HX	heat exchanger
IML	International Microgravity Laboratory
ISS	International Space Station
JSC	NASA Johnson Space Center, Houston, Texas
LMLSTP	Lunar Mars Life Support Test Project
LOAEL	lowest-observed-adverse-effect level
LSM	life support module (Russian segment of ISS)
MCL	maximum contaminant level
MCV	microbial check valve
MOA	mode of action
MSFC	Marshall Space Flight Center, Huntsville, Alabama
NASA	National Aeronautics and Space Administration
NOAEL	no-observed-adverse-effect level

NRC	National Research Council
PBPK	physiologically based pharmacokinetics
PCWQM	process-control water-quality monitor
PERP	permitted exposure/rodent potency
QD	quick disconnect
SL-J	Spacelab-J
SLS-1	Spacelab Life Sciences-Mission 1
SM	service module (Russian segment of ISS)
SMAC	spacecraft maximum allowable concentration
SRV-K	Russian condensate recovery system
SRV-U	Russian urine recovery assembly
SSP	Space Station Program
STS	space transportation system (refers to shuttle mission)
SWEG	spacecraft water exposure guideline
TIC	total inorganic carbon
TIMES	thermally integrated membrane evaporation system
TOC	total organic carbon
TOCA	total organic carbon analyzer
USML	U.S. Microgravity Laboratory
VCD	vapor compression distillation
VRA	volatile removal assembly
WRS	water recovery system
WRT	Water Recovery Test

Methods for Developing Spacecraft Water Exposure Guidelines

Executive Summary

TRAVEL, exploration, and study in space are challenging and fascinating scientific objectives for the 21st century. To be successful in those endeavors, the National Aeronautics and Space Administration (NASA) must continue to develop life support and technology programs for more frequent, complex, and longer missions. Because of the closed environment of spacecraft, an important issue is the inevitable accumulation of contaminants in the air and water systems. To prevent adverse health effects and degradation of work performance, it will be necessary to minimize space crews' exposures to those contaminants.

NASA has established exposure guidelines for airborne contaminants, called spacecraft maximum allowable concentrations (SMACs). SMACs are determined using guidelines developed for NASA by the National Research Council (NRC). However, for water contaminants, NASA's requirements have been based on standards from the U.S. Public Health Service and the U.S. Environmental Protection Agency for public drinking water. Those standards were established to protect the general public and are not always appropriate for application to NASA missions, because exposure conditions in space are different from those on Earth.

NASA requested that the NRC develop guidelines for setting exposure guidance levels for spacecraft water contaminants, similar to those established for airborne contaminants by the NRC in 1992 (*Guidelines for Developing Spacecraft Maximum Allowable Concentrations for Space Station Contaminants*). The NRC was asked to consider only chemical contaminants, not microbial agents. The NRC assigned this task to the Committee on Toxicology. The Subcommittee on Spacecraft Water Exposure Guidelines, a multidisciplinary group of experts, was convened to develop the guidelines presented in this report for calculating exposure levels that will prevent adverse health effects and degradation in crew performance. These guidance levels are called spacecraft water exposure guidelines (SWEGs).

SWEGs are to be established for exposures of 1, 10, 100, and 1000 days. The 1-day SWEG is a concentration of a substance in water that is judged to be acceptable for the performance of specific tasks during rare emergency conditions lasting for periods up to 24 hours. The 1-day SWEG is intended to prevent irreversible harm and degradation in crew performance. Temporary discomfort is permissible as long as there is no effect on judgment, performance, or ability to respond to an emergency. Longer-term SWEGs are intended to prevent adverse health effects (either immediate or delayed) and degradation in crew performance that could result from continuous exposure in closed spacecraft for as long as 1000 days. In contrast with the 1-day SWEG, longer-term SWEGs are intended to provide guidance for exposure under what is expected to be normal operating conditions in spacecraft.

The subcommittee used the NRC's 1992 SMAC guidelines as a general model for developing SWEGs. In addition, the subcommittee considered the following: (1) sources of spacecraft water contaminants, (2) methods to rank the contaminants for risk assessment, (3) relevance of available animal toxicity data for predicting toxicity to humans in space, (4) risk assessment methods for deriving exposure guidelines, (5) methods for modifying risk estimates to account for altered physiologic changes and stresses caused by microgravity, and (6) exposure guidelines established by other organizations.

WATER CONTAMINANT SOURCES

To provide a space crew with an adequate water supply, it is necessary to recycle spacecraft wastewater during long space flights. Water is needed for drinking, hygienic uses, and oxygen generation. Water on long space flights can be recovered onboard from several sources, including humidity condensate, used hygiene water, and urine. Humidity condensate will likely have the greatest contaminant variability because it will include contaminants released into the cabin from by-products of crew metabolism, food preparation, and hygiene activities; from routine operation of the air revitalization system; from off-gassing of materials and hardware; from payload experiments; from routine in-flight use of the crew health care system; and from other sources. Wash water will include detergents and other personal hygiene products. Urine contains electrolytes, small molecular weight proteins, and metabolites of nutrients and drugs. The urine is chemically treated and distilled before recycling, which causes a variety of by-products to be formed. Other sources of chemical contaminants include mechanical leaks, microbial metabolites, payload chemicals, biocidal agents added to the water to retard bacterial growth, fouling of the filtration system, and incomplete processing of the water.

Contaminants in the atmosphere can also end up as toxic substances in the water system. The air and water systems of the International Space Station constitute a single life support system, and the use of condensate from inside the cabin as a source of drinking water could introduce unwanted substances into the water system.

RANKING CONTAMINANTS FOR RISK ASSESSMENT

Ideally, SWEGs should be established for all chemical substances that might be found in spacecraft water. As a practical matter, it would be difficult to develop SWEGs for the more than 400 chemical species that have been identified on space missions in the past. Priorities are needed among the candidate chemicals for risk assessment. Setting priorities for risk assessment is a function separate from conducting the risk assessments themselves. There are three alternative approaches that NASA can use to select candidate chemicals for risk assessment. One involves a subjective selection of chemicals, in which selection parameters may not necessarily be specified. In this approach, NASA would make qualitative judgments about which chemicals to evaluate. The second approach would be to specify a set of parameters that should be considered when making a selection (such as the magnitude of routine and accidental exposures, short- and long-term effects, and ability to monitor and control exposure). A third approach would expand on the second by quantifying and weighing parameters and using a formula to calculate priorities for different substances. Each approach has benefits and limitations, and a successful system for selecting a substance might involve a combination of them.

DATA FOR ESTABLISHING SWEGS

In developing SWEGs, several types of data should be evaluated, including data on (1) the physical and chemical characteristics of the contaminant, (2) in vitro toxicity studies, (3) toxicokinetic studies, (4) animal toxicity studies conducted over a range of exposure durations, (5) genotoxicity studies, (6) carcinogenicity bioassays, (7) human clinical and epidemiology studies, and (8) mechanistic studies. All observed toxic effects should be considered, including mortality, morbidity, functional impairment, neurotoxicity, immunotoxicity, reproductive toxicity, genotoxicity, and carcinogenicity.

Data from oral exposure studies should be used, particularly drinking water and feed studies, in which the duration of exposure approximates anticipated human exposure times. Gavage studies can also be used, but they should be interpreted carefully because they involve the bolus administration of a substance directly to the stomach within a brief period of time. Such exposure could induce blood concentrations of contaminants and attendant effects that might not be observed if the administration were spread out over several smaller doses, as would be expected with the normal pattern of water consumption. Dermal absorption and inhalation studies should also be evaluated, because exposure from those routes occur when water is used for hygiene purposes.

There are several important determinants for deriving a SWEG, including identifying the most sensitive target organ or body system affected; the nature of the effect on the target tissue; dose-response relationships for the target tissue; the rate of recovery; the nature and severity of the injury; cumulative effects; toxicokinetic data; interactions with other chemicals; and the effects of microgravity.

RISK ASSESSMENT

There are several risk assessment methods that can be used to derive SWEGs. Risk assessments for exposure to noncarcinogenic substances traditionally have been based on the premise that an adverse health effect will not occur below a specific threshold exposure. Given this assumption, an exposure guidance level can be established by dividing the no-observed-adverse-effect level (NOAEL) or the lowest-observed-adverse-effect level (LOAEL) by an appropriate set of uncertainty factors. This method requires making judgements about the critical toxicity end point relevant to a human in space, the appropriate study for selecting a NOAEL or LOAEL, and the magnitudes of the uncertainty factors used in the process.

For carcinogenic effects known to result from direct mutagenic events, no threshold dose would be assumed. However, when carcinogenesis results from nongenotoxic mechanisms, a threshold dose can be considered. Estimation of carcinogenic risk involves fitting mathematical models to experimental data and extrapolating to predict risks at doses that are usually well below the experimental range. The multistage model of Armitage and Doll is used most frequently for low-dose extrapolation. According to multistage theory, a malignant cancer cell develops from a single stem cell as a result of several biologic events (e.g., mutations) that must occur in a specific order. There also is a two-stage model that explicitly provides for tissue growth and cell kinetics.

An alternative to the traditional carcinogenic and noncarcinogenic risk assessment methods is the benchmark-dose (BMD) approach. The BMD is the dose associated with a specified low level of excess health risk, generally in the risk range of 1%-10% (BMD_{01} and BMD_{10}), that can be estimated from modeled data with little or no extrapolation outside the experimental dose range. Like the NOAEL and LOAEL, respectively, the BMD_{01} and BMD_{10} are starting points for establishing exposure guidelines and should be modified by appropriate exposure conversions and uncertainty factors.

Scientific judgment is often a critical, overriding factor in applying the methods described above. The subcommittee recommends that when sufficient dose-response data are available, the BMD approach be used to calculate exposure guidelines. However, in the absence of sufficient data, or when special circumstances dictate, the other, more traditional approaches should be used.

SPECIAL CONSIDERATIONS FOR NASA

When deriving SWEGs, either by the traditional or BMD approach, it will be necessary to use exposure conversions and uncertainty factors to adjust for weaknesses or uncertainties about the data. When adequate data are available, exposure conversions that NASA should use include those to adjust for target tissue dose, differences in exposure duration, species differences, and differences in routes of exposure. Uncertainty factors should also be used to extrapolate animal exposure data to humans, when human exposure data are unavailable or inadequate; to extrapolate data from subchronic studies to chronic exposure; to account for using BMD_{10} instead of BMD_{01} (or a LOAEL instead of a NOAEL); to account for experimental variation; and to adjust for space-flight factors that could alter the toxicity of water contaminants. The latter factors are used to account for uncertainties associated with microgravity, radiation, and stress. Some of the ways astronauts can be physically, physiologically, and psychologically compromised include decreased muscle mass, decreased bone mass, decreased red blood cell mass, depressed immune systems, altered nutritional requirements, behavioral changes, shift of body fluids, altered blood flow, altered hormonal status, altered enzyme concentrations, increased sensitization to cardiac arrhythmia, and altered drug metabolism. There is generally little information to permit a quantitative conversion that would reflect altered toxicity resulting from spaceflight environmental factors. Thus, space-flight uncertainty factors should be used when available information on a substance indicates that it could compound one or more aspects of an astronaut's condition that might already be compromised in space.

Another commonly used uncertainty factor is one that accounts for variable susceptibilities in the human population. That uncertainty factor is used to protect sensitive members of the general population, including young children, pregnant women, and the immune compromised. Because the astronaut population is typically composed of healthy nonpregnant adults, the subcommittee believes that an uncertainty factor for intraspecies differences should only be used if there is evidence that some individuals could be especially susceptible to the contaminant.

EXPOSURE GUIDELINES SET BY OTHER ORGANIZATIONS

Several regulatory agencies have established exposure guidance levels for some of the contaminants of concern to NASA. Those guidance levels should be reviewed before SWEGs are established. The purpose of this comparison would not be simply to mimic the regulatory guidelines set elsewhere, but to determine how and why other exposure guidelines might differ from those of NASA and to assess whether those differences are reasonable in light of NASA's special needs.

1

Introduction

CONSTRUCTION of the International Space Station (ISS)—a multinational effort—began in 1999. In its present configuration, the ISS is expected to carry a crew of three to six astronauts for up to 180 days. Because the space station will be a closed and complex environment, some contamination of its internal atmosphere and water system is unavoidable. Several hundred chemical contaminants are likely to be found in the closed-loop atmosphere and recycled water of the space station.

In 1992, the NRC provided NASA with guidelines for developing exposure guidance levels for airborne contaminants in *Guidelines for Developing Spacecraft Maximum Allowable Concentrations* (SMACs) *for Space Station Contaminants* (NRC 1992). That report provides guidance on the sources and types of data that should be used for establishing SMACs, approaches for performing risk assessments for carcinogenic and noncarcinogenic effects, and how to account for the effects of physiological changes induced by microgravity. SMACs have been established for 50 airborne contaminants using the NRC guidelines (NRC 1994; 1996a,b; 2000).

To protect space crews from contaminants in potable and hygiene water, NASA requested that the NRC develop guidelines, similar to those established by the NRC for airborne contaminants, for setting exposure guidance levels for spacecraft water contaminants. The NRC was asked to consider only chemical contaminants, and not microbial agents. The NRC assigned this task to the Committee on Toxicology, and the Subcommittee on Spacecraft Water Exposure Guidelines, a multidisciplinary group of experts, was convened to develop guidelines for calculating exposure levels that will prevent adverse health effects and degradation in crew performance. These guidance levels are called spacecraft water exposure guidelines (SWEGs).

SWEGs are to be established for exposures of 1, 10, 100, and 1000 days. The 1-day SWEG is a concentration of a substance in water that is judged to be acceptable for the performance of specific tasks during rare emergency conditions lasting for periods up to 24 hours. The 1-day SWEG is intended to prevent irreversible harm and degradation in crew performance. Temporary discomfort is permissible as long as there is no effect on judgment, performance, or ability to respond to an emergency. Longer-term SWEGs are intended to prevent adverse health effects (either immediate or delayed) and degradation in crew performance that could result from continuous exposure in closed spacecraft for as long as 1000 days. In contrast with the 1-day SWEG, longer-term SWEGs are intended to provide guidance for exposure under what is expected to be normal operating conditions in spacecraft.

SWEGs and SMACs differ from each other in two fundamental ways. The first is that SMACs are used for inhalation exposures, whereas SWEGs will be used for oral exposures. Second, the time scales used to set the guidance levels are different. SMACs were developed for 1-hr, 24-hr, 7-day, 30-day, and 180-day exposures, whereas SWEGs will be established for 1, 10, 100, and 1000 days. The reason for the difference is that exposure to water is more intermittent than is exposure to air and because it would be possible to refrain from drinking or using contaminated water for short periods in an emergency. Furthermore, there is a possibility that NASA could conduct missions that would last for up to 1000 days, so a long-term SWEG is needed.

WATER CONTAMINANTS

Water used in NASA's space missions must be carried from Earth or generated by fuel cells. The water is used for drinking, food reconstitution, oral hygiene, hygienic uses (handwashing, showers, urine flushing), and oxygen generation. Because of plans for longer space flights and habitation of the ISS, water reclamation, treatment, and recycling is required. Water for long space flights can be reclaimed from several onboard sources, including humidity condensate from the cabin, hygiene water (shower and wash water), and urine. Each of those sources will have a variety of contaminants. Humidity condensate will have contaminants released into the cabin from crew activities (e.g., by-products of crew metabolism, food preparation, and hygiene activities); from routine operation of the air revitalization system; from off-gassing of materials and hardware; from payload experiments; and from routine in-flight use of the crew health care system. Wash water will include detergents and other personal hygiene products. Urine contains electrolytes, small molecular weight proteins, and metabolites of nutrients and drugs. The urine is chemically treated and distilled before recycling, which causes a variety of by-products to be formed. Other sources of chemical contaminants include mechanical leaks, microbial metabolites, payload chemicals, biocidal agents added to the water to retard bacterial growth (e.g., silver, iodine), fouling of the filtration system, and incomplete processing of the water.

The possibility also exists that contaminants in the atmosphere can end up as toxic substances in the water system. The air and water systems of the ISS constitute a single life support system, and the use of condensate from inside the cabin as a source of drinking water could introduce some unwanted substances into the water system.

NASA's current water exposure guidelines are based on standards from the U.S. Public Health Service and the U.S. Environmental Protection Agency (EPA) for public drinking water. Those standards were established to protect the general public, including the elderly, persons with disabilities or compromised immune systems, and infants. Protecting sensitive individuals is necessary and appropriate for the safety of the public health, given the likelihood of lifetime exposures. However, exposure limits for the general public are not necessarily appropriate for spacecraft flight crews. Many of the limits are likely to be overly conservative—much stricter than would be necessary to protect healthy adult astronauts even for several years away from Earth. Other limits will be inadequate—microgravity, increased radiation, or other unique attributes of spaceflight could make astronauts more sensitive than are humans on Earth to a given contaminant. Moreover, water exposure guidance levels are not available for many contaminants that might be found in spacecraft water supplies. Data collected from space shuttle and *Mir* missions indicate that organic compounds found in processed water samples are vastly different from the list of target compounds developed by EPA for protection of public drinking-water supplies.

APPROACH TO THE STUDY

NASA briefed the subcommittee on the water reclamation systems for the ISS and its programmatic predecessors, and provided water-contaminant data collected from ground-based and in-flight tests. That information is presented in Chapter 2, and also includes a description of the treatment methods and monitoring strategies for spacecraft water. Using those data and the SMACs guidelines, the subcommittee considered the sources and types of data that should be used in establishing SWEGs. That assessment is provided in Chapter 3, with particular emphasis given to advancements made since the establishment of the SMACs guidelines in the areas of neurobehavioral toxicity, reproductive toxicity, mutagenicity, epidemiology, and toxicologic mechanisms. The subcommittee also considered available approaches for establishing exposure guidelines for waterborne contaminants. A review of those approaches and the subcommittee's recommendations for deriving SWEGs in provided in Chapter 4. That chapter also discusses ways to account for uncertainties associated with spaceflight, such as microgravity. Because it is not possible to conduct risk assessments on all the potential water contaminants, the subcommittee also considered prioritizing the contaminants for risk assessment. Approaches to ranking spacecraft water contaminants are provided in Chapter 5.

REFERENCES

NRC (National Research Council). 1992. Guidelines for Developing Spacecraft Maximum Allowable Concentrations for Space Station Contaminants. Washington, DC: National Academy Press.

NRC (National Research Council). 1994. Spacecraft Maximum Allowable

Concentrations for Selected Airborne Contaminants, Volume 1. Washington, DC: National Academy Press.
NRC (National Research Council). 1996a. Spacecraft Maximum Allowable Concentrations for Selected Airborne Contaminants, Volume 2. Washington, DC: National Academy Press.
NRC (National Research Council). 1996b. Spacecraft Maximum Allowable Concentrations for Selected Airborne Contaminants, Volume 3. Washington, DC: National Academy Press.
NRC (National Research Council). 2000. Spacecraft Maximum Allowable Concentrations for Selected Airborne Contaminants, Volume 4. Washington, DC: National Academy Press.

2

Sources, Treatment, and Monitoring Of Spacecraft Water Contaminants

THIS chapter provides an overview of the water reclamation system of the International Space Station (ISS) and a discussion of sources of spacecraft water contamination. Water treatments are discussed in conjunction with contaminant sources, as the treatments also contribute to the contamination. Strategies for monitoring water quality are also discussed.

OVERVIEW

The ISS is expected to operate for many years, with each crew spending up to 6 months onboard. The prohibitive cost of transporting the large amounts of water needed to support the crew and the impracticality of generating water from fuel cells for missions of this length have led to the requirement that the ISS environmental control and life support system (ECLSS) recycle wastewater to provide water of acceptable quality for potable and personal hygiene use and for oxygen generation.

In 1992, the design team for the U.S. space-station life-support systems was directed to assess existing Russian technologies for possible use in developing life support hardware for the ISS (Mitchell et al. 1994). Many components of Russian life support systems (e.g., atmosphere revitalization and water recovery) already have been operational in microgravity. A means of recovering water from humidity condensate has been in use since the *Salyut* era of the mid-1970s, and a urine-processing system has operated on the *Mir* station since 1989.

The life support system for the ISS will be incorporated in several phases of the assembly sequence. Initially, potable water will be produced from humidity condensate by multifiltration treatment by a Russian assembly housed in the Russian service module. Fuel cell water from the U.S. space shuttle will be transferred to the station after docking. This water, stored in special tanks on the station, will provide an emergency supply. The plan will accommodate up to three crew members. An advanced Russian life support system, involving hygiene-water processing and urine processing will be deployed in the Russian life support module (LSM) of the ISS. At approximately the same time, the ability to reclaim potable water from humidity condensate, hygiene wastewater, and urine distillates (via multifiltration and catalytic oxidation) will be incorporated in the U.S. habitation module of the ISS.

Design Drivers

Water reclamation systems intended for the ISS and its programmatic predecessors have been designed to deliver specific amounts of product water (Table 2-1) of specified quality. (At the time the ISS system design was begun, no models were available to predict contaminant loads in water produced by the water reclamation systems. Since then, considerable information has been generated from ground-based and in-flight studies, as described later in this chapter, and a predictive model is being generated (D.L. Carter, Marshall Space Flight Center, personal communication, Oct. 13, 1999).) Other considerations in designing the water reclamation systems included shelf life, resupply-return logistics, crew time needed for maintenance, power needed to operate the system, launch weight, and stowage volume. The processing assemblies for the ISS have been designed to support six crew members after assembly is complete. The product-water tank's capacity will be about 120 lb of processed water.

TABLE 2-1 ISS Water Requirements

Purpose	Amount, lb	Amount, kg
Drinking, food rehydration, oral hygiene	6.2/person/d	2.81/person/d
Extravehicular activity	20	9.05
High per-person usage	11.35 over 24 hr	5.14 over 24 hr
Hygiene	15.0/person/d	6.79/person/d
Hygiene (high usage)	16.0/person over 24 hr	7.24/person over 24 hr
Life sciences experiments (with animals)	7.35/d	3.33/d
Maximum off-line water-quality analysis	2.2/d	1.00/d
Nominal off-line water-quality analysis	1.7/d	0.77/d
Oxygen generation	17/d	7.69/d
Payload experiments	4.8/experiment d	2.17/experiment d

The ISS is designed to support six crew members.

Source: Segment Specification for the U.S. On-Orbit, Specification Number SSP 41162E, July 1996, p. 273.

Table 2-2 illustrates the water mass balance of the Russian water-processing segments before the completion of the ISS. Total water consumption of a cosmonaut was estimated at approximately 9 lb or 4.1 liters per day (L/d). During the first phase of construction, the Russian service module will reclaim drinking water only from humidity condensate; during the second phase, the service module and the LSM will regenerate potable water from urine. During the second phase, the amount of water supplied by the Russian progress vehicle and the U.S. space shuttle needed to make up the water balance will be lower.

ISS Water-Quality Standards

Standards for recycled water in a closed spacecraft system have been a matter of debate for many years. Because water aboard the U.S. space shuttle is not recycled (it is generated by onboard fuel cells), ex-

TABLE 2-2 Water Mass Balance Estimates of the Russian Water-Processing Segments

Water Demand (lb/person/d)		Water Supply Sources, lb Source	First Phase (SM)	Second Phase (SM + LSM)
Drinking, food	5.5	Humidity condensate	3.3	3.3
Oxygen generation by electrolysis	2.21	Hygiene water evaporation	0.66	0.66
Personal hygiene	1.1	Water with food	1.1	1.1
Urinal flush	0.66/0.22	Water from storage system	4.44	0.44
		Water from WRS-UM	NA	2.58
		Water from CDRS	NA	0.95
Total	9.48/9.0		9.48	9.04

Water mass balance estimates did not consider evaporation or other water losses. 85% of the crew water will be regenerated during phase 2, whereas only 43% during phase 1. CDRS, carbon dioxide reduction system; LSM, life support module; NA, not applicable; SM, service module; WRS-UM, updated system for water reclamation from urine

Source: Modified from Samsonov et al. (1997). The units have been converted to pounds for comparison with Table 2-1.

isting water-quality standards for shuttle water cannot be extended to recycled water, particularly for long ISS missions. In 1986 and 1989, the National Research Council (NRC) Committee on Toxicology reviewed water-quality standards for the National Aeronautics and Space Administration (NASA), and recommend maximum contaminant levels that would protect the health of crews on long space missions (NRC 1986, 1989). This information was used in the design of onboard water treatment and recycling system for the ISS. One critical recommendation was that the integrated ECLSS should be able to assess possible interactions between the air revitalization system and the water reclamation system in terms of contaminants that might arise from one or the other system.

The current requirements for ISS water quality are based on standards from the U.S. Public Health Service (PHS 1962) and U.S. Environmental Protection Agency (EPA 1996) for public drinking water. Standards for water quality are described in Appendix A and in other NASA documents (e.g., the International Space Station Flight Crew Integration Standard, SSP 50005, Rev B, Aug. 1995).

Because the Russian segments will initially support water regeneration for the ISS, a review and consensus on Russian water-quality standards is necessary. The standards established for the *Mir* station generally are less stringent than are U.S. standards: Fewer limits are specified, and those that are specified tend to have higher maximum limits. Also, the disinfectant used for the *Mir* water-processing system is silver, whereas the biocide for most U.S. systems is iodine. Differences between the two programs in analytical techniques have made direct comparisons difficult; nevertheless, negotiations are currently under way to develop water-quality standards that are acceptable to both partners in the ISS program. Appendix A, Table A-2, compares the proposed standards—which have yet to be accepted officially—of the Russian and U.S. programs.

Some detailed descriptions of the water reclamation systems that are currently operating on *Mir*, planned upgrades for the Russian segments of the ISS, and the planned U.S. integrated water reclamation system functions are provided in Appendix A.

SOURCES OF SPACECRAFT WATER CONTAMINATION

The water reclamation system for the ISS comprises a unique combination of input and output streams. Waste streams will include urine and urine flush water, humidity condensate, personal hygiene water (body wash), water from general hygiene activities (hand washing, shaving, teeth cleaning), and effluent from the crew health care system (CHeCS). Humidity condensate undoubtedly will have the greatest inherent variability, because it will include contributions from crew metabolism, hygiene activities, food preparation, materials off-gassing, and payload experiments, some of which will involve animals and all of which will vary widely from mission to mission.

The nature of the water sources and the extent of closure in the recycling loop have posed substantial challenges for defining product-water specifications that will protect crew health over long periods. In its earlier reviews, NRC (1986, 1989) evaluated NASA's plans for potable-water reclamation, including issues on the small volume of water; the complexity and variability of cabin humidity condensate (the principal source of potable water); the tight interface between cabin air and water systems; the treatment processes for urine and hygiene wastewater and the treatment by-products; the potential accumulation of polar, uncharged organic molecules; the need for toxicologic characterization of unique chemical and microbial by-products; and monitoring and analytical capabilities. Since those reviews, considerable progress has been made in the characterization of source-water contaminants.

The results described here come from ground-based and in-flight studies (from *Spacelab*, shuttle, and *Mir* missions). Two ground-based test beds in particular have generated a wealth of information on humidity condensate, urine distillates, urine off-gassing products, and wash water—the Water Recovery Test (WRT) at NASA's Marshall Space Flight Center in Huntsville, Alabama, and the Early Human Testing Initiative (renamed the Lunar-Mars Life Support Test Project (LMLSTP)) at NASA's Johnson Space Center in Houston, Texas. The Marshall test bed has been used mainly to assess the performance of the various systems intended for environmental control and life support on space stations. Stages 9 and 10, the two most recent versions of this test bed simulate many aspects of the configuration planned for the ISS. Figure 2-1 is a diagram of the Stage 9 water recovery system. Humans participate in the operation of this system through brief visits to the end-user equipment facility, which includes a shower, hand-washing sink, microwave oven, urine collection and pretreatment unit, condensing heat exchanger, and exercise equipment.

The LMLSTP, in contrast, involves human subjects actually living within a closed test chamber that has integrated air and water reclamation systems. Subjects in the chamber donate and use recycled water. Three tests have been completed with this system, the first a 15-d mission involving one crew member, and the second and third involving 30- and 60-d missions, respectively, each including a four-person crew. Results from a 90-d test are not discussed here because water reclamation processes other than those planned for ISS were used.

Some information about flight data is available from the 1985 U.S. *Spacelab-3* mission; more recent data were generated through observational studies begun in the 1990s on the space shuttle. The advent of

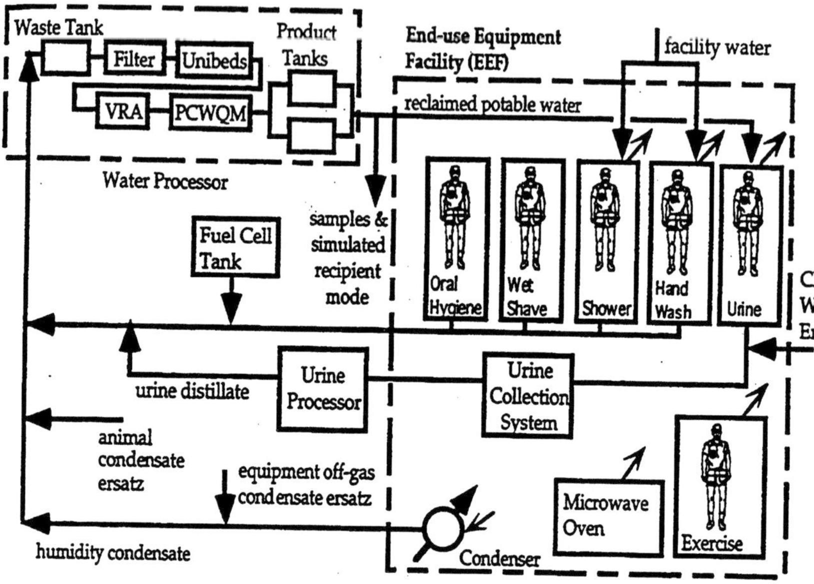

FIGURE 2-1 The integrated water recovery system used in Stage 9 of the Marshall WRT. CHeCS, crew health-care system; PCWQM, process-control water-quality monitor; VRA, volatile removal assembly. Source: Holder et al. (1995).

U.S.-Russian cooperation has provided an invaluable opportunity to assess the *Mir* spacecraft water-recycling system, which operates in microgravity. Results from these and other studies form the existing database of the likely chemical constituents of the proposed ISS water system.

The major sources of water and their likely contaminants are discussed below. These include likely constituents of human urine, both before and after treatment; humidity condensates and wash water, and the various sources of chemical contaminants to those streams; and chemical by-products that can form as a result of treatment failures or from the use of biocides.

Human Urine

Untreated Urine

The composition of human urine is extremely complex, and it varies widely according to diet, use of medications, and health status. The normal constituents of urine include electrolytes, small-molecular-weight proteins, and metabolites of nutrients and drugs. Typical components include salts of iron, sodium, potassium, magnesium, calcium, chlorides, phosphates, citrates, oxalates, and sulfates. Also present are hormones, vitamins, creatinine, uric acid, and carboxylic acids.

Crews in the *Apollo*, *Skylab*, and space-shuttle programs have taken drugs to alleviate motion sickness, headache, sleeplessness, constipation, and nasal congestion during flights (Putcha et al. 1994). Because those drugs are likely to be used aboard the ISS, their metabolites will be introduced into the water system via urine. As a result, the chemical composition of urine in the ISS might be somewhat different from the urine used to develop and test ground-based systems where such medications have not been incorporated.

Chemical Treatment and Distillation By-Products

The water reclamation system planned for the ISS uses chemical pretreatment and distillation in treating urine. Oxone (a commercially developed potassium monopersulfate compound) and sulfuric acid are added to the raw urine to stabilize it, to fix ammoniated species, and to control microbial content. Oxone can oxidize chloride compounds in the urine to form chlorine, which will react directly with several organic compounds to form chlorinated hydrocarbons. This process can generate several nonphysiologic chlorination by-products. Cole et al. (1991) have reported off-gassing of cyanogen chloride, chlorinated ketones, and chlorinated nitriles from treated urine, which probably appeared because of high ammonia and amino acid content in the raw urine.

Several distillation technologies have been used in developing the water reclamation system for the space station (including reverse osmosis, thermally integrated membrane evaporation (TIMES), or vapor compression distillation). Early stages of the WRT included TIMES distillation, but later versions included vapor compression distillation, which is the current technology of choice for the ISS. (Urine was pretreated with Oxone and sulfuric acid in both distillation processes.) Despite some minor differences in the composition of the distillate from the two processes, the contaminant load (and its variations over time) were thought to reflect those of the space-station system (Winkler et al. 1983; Verostko 1986; Cole et al. 1991; Carter and Bagdigian 1993). The TIMES distillate included carboxylates, alcohols, ketones, aldehydes, phenols, nitriles, hydrocarbons, and halogenated hydrocarbons. Most compounds could be traced to human metabolism, but some seemed to have resulted from the urine pretreatment process. The constituents produced by vapor compression distillation in Stages 4 and 5 of the WRT are described by Carter et al. (1992).

Vapor compression distillation also was used to treat urine in the closed-chamber tests at the Johnson Space Center. Compounds found in the urine distillate samples from the 30-d test included: acetic, butyric, formic, propionic, and lactic acids; ethanol; methanol; 1,1-dichloropropanone; and ibuprofen. Formic acid was present in the highest concentration (13,700 micrograms per liter (μg/L)) (Homan et al. 1997; Verostko et al. 1997).

In addition to these results, for the raw distillate of the urine, additional information is available concerning the volatile oxidation products arising from urine pretreatment. In 1989, the NRC expressed serious concerns about concentrations of cyanogen chloride (100 parts per billion (ppb)), reportedly off-gassed in the Marshall WRT study. Off-gassed products from urine treated with Oxone and sulfuric acid (the treatment chosen for the ISS) during a Marshall WRT have been described by Cole et al. (1991). High concentrations of acetone, acetonitrile, methylene chloride, 3-methylbutanal, dimethylamine, propanenitrile, and several other volatile oxidation products were reported. More than a dozen compounds occurred frequently in the off-gassing products from the seven urine samples analyzed.

Humidity Condensate

Humidity condensate will be an important source water—and probably the most variable—in the ISS water reclamation system. Humidity condensate is collected by the cabin heat exchanger, which is controlled by the spacecraft's air revitalization system. The chemical constituents will include contaminants released into the cabin air from crew activities, such as by-products of metabolism, food preparation, and hygiene activities (including the use of cleansers and disinfectants); from routine operation of the air revitalization system; from materials and hardware off-gassing; from payload experiments, especially those involving animals; from onboard utility chemicals; and from routine in-flight use of the CHeCS. Some steps in the water recovery process involve liquid-air separations. The air from those steps is vented into the cabin atmosphere and eventually appears in the humidity condensate. As wastewater passes through the multifiltration bed to the volatile removal assembly, for example, small molecular-weight organic compounds in the water undergo catalytic oxidation in the presence of sparged oxygen in a catalytic reactor. Before the effluent is passed to an ion exchange resin, which removes the small-molecular-weight organic compounds, the excess oxygen and other volatile compounds (incomplete oxidation products) are vented to the cabin. Even though some of these airborne contaminants will be removed by "scrubbers" in the air revitalization system, many will end up in the humidity condensate.

Designing an effective water treatment system requires that the chemical constituents of the source water, and the variability of those constituents, be thoroughly characterized. As a corollary, that system also must include ways to evaluate potential toxicologic hazards posed by consuming the product water. The information available regarding the chemical composition of spacecraft humidity condensates is described below. These results came from test bed studies (the WRT and the LMLSTP) and from postflight analyses of samples collected during actual spacecraft missions. The remainder of this section constitutes descriptions of other environmental contributors to condensate aboard spacecraft.

Condensate Sample Results

Water Recovery Test

The water recovery system of the Marshall Space Flight Center WRT includes a facility that contains a shower, handwash, microwave oven, urine collection and pretreatment unit, condensing heat exchanger, and exercise equipment. Tests in which human subjects use this facility have revealed the presence of low-molecular-weight acids, semivolatile acids, volatile alcohols, purgeable organic compounds, semivolatile organic compounds, and glycols in the humidity condensate (Cole et al. 1991). The chemical composition of the condensate was highly variable, no doubt reflecting variations both in the subjects' activities and in the ambient test environment.

Lunar-Mars Life Support Test Project

Part of the 30-d test of the Johnson Space Center's LMLSTP, in which wastewater was recycled throughout the test, included chemical characterization of humidity condensate samples. Results from four sets of samples collected are shown in Table 2-3. Total organic carbon (TOC) in the condensate ranged from 45 milligrams per liter (mg/L) to 65 mg/L, and accountability was 97%. Acetone, 2-butoxyethanol, diethyl phthalate, phenol, alcohols, and glycols were present in high concentrations, as were formaldehyde (7.7-12 mg/L) and ethylene glycol (5-12 mg/L). EPA's recommended lifetime maximum exposures are 1 mg/L for formaldehyde and 7 mg/L for ethylene glycol (EPA 1996).

Space Shuttle

Humidity condensate has been collected on relatively few space-shuttle missions, and most of the samples have been collected during the past 4 or 5 years. One exception was the 1985 STS-51B/*Spacelab-3* mission, which included tests with rodents housed in a new holding facility and analyses of the air revitalization system (Verostko 1986). Although the integrity of the samples collected from the mission was questionable, many organic compounds were identified, including alcohols, amides, amines, carboxylic acids, ethers, esters, ketones, phenols, and thiourea. In 1991 and 1992, samples were collected after four *Spacelab* missions (STS-40/SLS-1; STS-42/IML-1; STS-50/USML-1; STS-47/SL-J); however, the long delays after landing until samples could be collected undoubtedly affected their chemical composition. Finally, routine assessments of atmospheric quality aboard shuttle missions have revealed the presence of a wide spectrum of organic compounds that could well appear in the humidity condensate (James et al. 1994).

The first in-flight humidity condensate samples were collected on two shuttle missions, STS-45 and STS-47, in 1992; additional samples were collected from STS-68 and from the *Mir*. Organic compounds found in the 9 samples collected on the STS-45 and STS-47 missions are listed in Table 2-4 (Muckle et al. 1993).

TABLE 2-3 Organic Constituents of Humidity Condensate from the LMLSTP 30-d Closed-Chamber Test

Compound	Maximum Concentration (µg/L)
Volatile Organic Compounds	
Acetone[a]	2170
2-Butanone	91.71
Cyclohexanone	111.11
Methylene chloride	1.78
4-Methyl-2-pentanone	14.12
Tetrahydrofuran	25.22
Toluene	1.26
Extractable Organic Compounds	
Acetophenone	56.3
2-Acetylfuran	1.1
Benzaldehyde	3.8
Benzoic acid	29.9
Benzothiazole	32.7
Benzyl alcohol	21.1
Benzylbutyl phthlate	4.9
Benzyl salicylate	1.4
Bis-2-ethylhexyl adipate	2.1
Bis-2-ethylhexyl phthalate	13.2
2-Butoxyethanol[a]	3533.8
2-(2-Butoxyethoxy)ethanol	136.6
2-(2-Butoxyethoxy)ethanol acetate	5.7
N-Butyl palmitate	2.5
Butylated hydroxyanisole (BHA)	11.7
n-Butylbenzenesulfonamide	1.0
4,4'-Butylidenebis (6-*tert*-butyl-*m*-cresol)	5.7
3-*tert*-Butylphenol[a]	125.3
Camphor	1.3
Caprolactam	60.1
Carvone	13.4
trans-Cinnamaldehyde	0.9
Cyclohexanol	79.8
Decamethylcyclopentasiloxane	1.2
Decanoic acid	36.2
1,4-Diacetylbenzene	36.9
Di-*n*-butyl phthlate	23.9
Di-*n*-butylamine	10.8
N,N,-Dibutyl formamide	23.5
3,5-Di-*t*-butyl-4-hydroxy-benzaldehyde	3.1
2,6-Di-*t*-butyl-4-methylphenol	3
2,4-Di-*t*-butylphenol	1.7
Diethyl phthalate[a]	214.9
Diethylene glycol monoethyl ether	137.4
N,N-Diethyl formamide	28.3
N,N-Diethyl-*m*-toluamide	2.2
Diisobutyl phthlate	3.7
Diisopropyl adipate[a]	157.1
Dimethyl phthlate	98.9
N,N-Dimethylacetamide	47.7
N,N-Dimethylbenzylamine	0.3
N,N- Dimethylformamide[a]	311.3
2,3-Dimethylmaleic anhydride	1.2
2,4-Dimethylphenol	2.0
Dipropylene glycol methyl ether	43.7
Dodecamethylcyclohexasiloxane	1.0
Dodecanoic acid	20.6
2-Ethoxyethanol	8.7
Ethylene glycol monobutyl ether acetate	17.0
2-Ethylhexanoic acid[a]	327.1
2-Ethyl-1-hexanol[a]	130.9
Ethyl 4-hydroxybenzoate	2.5
4-Ethylmorpholine	8.9
4-Ethylphenol	3.6
Furfural	4.4
Furfuryl alcohol	4.7
Guaiacol (2-Methoxyphenol)	1.2
Heptanoic acid	30.1
2-Heptanone	15.9
1-Hexadecanol	12.1
1,6-Hexanediol	23.4
Hexanoic acid	118
2-Hexanol	1.6
γ-Hexanolactone	3.3
2-Hydroxybenzothiazole	3.3
3-Hydroxy-2-butanone	136.4
4-Hydroxy-4-methyl-2-pentanone	86.1
Indole	0.4
Isophorone	23.2
2-Isopropylphenol	8.2
3-Isopropylphenol	28.3
p-Menth-1-ene-8-ol	26.2
Mesityl oxide	1.0
3'-Methylacetophenone	3.5

TABLE 2-3 *(Continued)*

Compound	Maximum Concentration (μg/L)
3-Methyl-2-cyclohexen-1-one	2.2
Methyl-4-hydroxybenzoate	4.0
2-Methylphenol	2.7
4-Methylphenol	12.3
2-Methylpyrazaine	6.4
1-Methyl-2-pyrrolidinone	39
Methyl stearate	0.4
Methyl sulfone	6.6
2-Methylthiobenzothiazole	0.4
Monomethyl phthalate	89.7
Myristic acid	6.2
Neomenthol	83.0
Nicotine	2.1
2-Nitrophenol	0.6
Nonanoic acid	55.9
Octadecanol	23.8
Octanoic acid	65.5
4-*tert*-Octylphenol	1.3
Pentacosane	0.9
Pentanoic acid	50.1
2-Pentanone	1.9
sec-Phenethyl alcohol	3.0
Phenol[a]	427.5
2-Phenoxyethanol	43.6
1-Phenoxy-2-propanol	9.3
2-Phenylacetic acid	12.9
Phenylethyl alcohol	6.5
2-Phenylphenol	0.8
2-Phenyl-2-propanol[a]	347.5
Phthalide	7.3
2-Propoxyethanol	6.1
Pulegone	5.9
Terpinen-4-ol	3.4
1-α-Terpineol	19.7
1-Tetradecanol	6.5
Tetramethylsuccinonitrile	61.0
Tetramethylthiourea	2.6
Tetramethylurea	6.1
Toluene	3.1
1,3,5-Triallyl-1,3,5-triazine-2,4,6(1*H*,3*H*,5*H*)-trione	4.5
Tributyl phosphate[a]	269.2
Tricosane	2.2
Triethylamine	27.7
Triethylene glycol methacrylate	1.7
Triethyl phosphate	4.6
2,2,4-Trimethyl-1,3-pentanediol[a]	268.8
2,2,4-Trimethyl-1,3-pentanediol diisobutyrate	21.7
Triphenyl phosphate	2.0
Tripropylene glycol monomethyl ether	109
Tris(2-chloroethyl) phosphate	2.6
Undecanoic acid	7.3
2-Undecanone	0.7
Valeraldehyde	1.8
γ-Valerolactone	1.8
Vanillin	1.3
Alcohols	
1-Butanol[a]	1480
Ethanol[a]	27,458
Methanol[a]	39,498
1-Propanol[a]	580
2-Propanol[a]	24,670
Glycols	
Ethylene glycol[a]	12,490
Propylene glycol[a]	8814
Aldehydes	
Formaldehyde[a]	12,192
Carboxylates	
Acetate[a]	9.21
Butyrate	0.21
Formate[a]	8.84
Lactate	0.183
Propionate[b]	0.32
Amines	
Trimethylamine	ND
Nonvolatile Compounds	
Urea	0.43
Total organic carbon	45,400 to 64,500

[a]Contaminant present in four of four samples collected.
[b]Contaminant present in three of four samples collected.
ND, target compound was not detected at or above the detection limit for that method.
Source: Homan et al. (1997).

TABLE 2-4 Organic Compounds Found in the Humidity Condensate Samples from the Shuttle Cabin

	Concentration, µg/L	
Compound	Minimum	Maximum
Volatile Organic Compounds		
Acetaldehyde	<10	19.8
Acetone	12.6	68
2-Heptanone	0.73	7.26
4-Heptanone	ND	6.03
Methylene chloride	1.94	1,380
Tetrachloroethane	2.67	10.60
Toluene	<0.25	1.4
Semivolatile Organic Compounds		
Acetophenone	22	108
2-(2-Butoxyethoxy)ethanol	2,700	3,400
3-*tert*-Butylphenol	36	193
Diethyl phthalate	480	2,200
N,*N*-Diethyl-*m*-toluamide	126	245
Octanoic acid	1,500	1,800
Phenol	35	107
1,3,5,Triazine-2,4,6(1*H*,3*H*,5*H*)-trione-1,3,5-tri-2-propenyl	480	2,300
Alcohols		
Benzylalcohol	59,000	369,000
Ethanol	6,600	126,000
Methanol	1,400	7,400
2-Propanol	1,900	43,300
Glycols		
Propylene glycol	29,000	72,000
Organic Acids		
Acetic acids	2,300	28,520
Butyric acid	280	900
Formic acids	8,900	16,900
Propionic acid	2,600	5,020
Aldehydes		
Formaldehyde	3,300	10,400

Representative results from the characterization of cabin humidity concentrations from nine samples from two shuttle missions. ND, not detected.

Source: Muckle et al. (1993).

Among the volatile organic compounds present, acetone, acetaldehyde, and methylene chloride were found in significant concentrations. Notably, most of the EPA target compounds were *not* found, and only methylene chloride exceeded the EPA maximum contaminant limit (MCL) of 0.005 mg/L for potable water. This finding underscores the unique nature of spacecraft source waters. Propylene glycol was the predominant glycol in the in-flight samples. Caprolactam, organic acids (acetic, formic, and propionic), and formaldehyde also were found, the latter in substantial amounts both in the cabin and in the *Spacelab* samples.

TOC accountability was about 50% for one mission and 75% for the other (Muckle et al. 1993); accountability in a similar study conducted on STS-68 was 86-88% (Straub et al. 1995). In the latter mission, 116 volatile organic compounds, 97 semivolatile organic compounds, caprolactam, urea, several organic acids, and formaldehyde were among the major contributors to the TOC. Alcohols (ethanol and 2-propanol) and glycols were the predominant organic alcohols. Variations in the amount of TOC among missions probably reflect both the shuttle vehicle and the type of payload launched. For example, the condensate taken from the shuttle with a *Spacelab* contained much higher TOC (Straub et al. 1995).

A discussion of inorganic contaminants found in humidity condensates from shuttle missions are presented later in this chapter, in conjunction with analyses of regenerated water.

Mir *Humidity Condensate*

In 1995 and 1996, another series of investigations was undertaken to characterize humidity condensate and processed water aboard *Mir*. Nine humidity condensate samples were collected, one during Mir-19, four during *Mir-20*, and four during *Mir-21*. (Samples from *Mir-20* and *Mir-21* included those collected during periods in which ethylene glycol leaked from the coolant system into the *Mir* atmosphere). Results from *Mir-19* and *Mir-20* are discussed separately from *Mir-21*, because advances in technology between the two study periods allowed closer examination of short-chain alcohols and urea in the *Mir-21* samples.

A wide spectrum of organic compounds was present in the *Mir-19* and *Mir-20* samples (Pierre et al. 1996). Differences were found between those samples and the earlier shuttle condensate samples in conductivity, turbidity, and anion and cation concentrations. Moreover, about 60 semivolatile compounds identified in the shuttle condensate were not found in *Mir* condensate, and about 10 compounds were found only in the *Mir* samples.

Volatile organic compounds common to the *Mir* and shuttle samples included acetone, 2-butanone, and methylene chloride. Others included acetaldehyde, carbon disulfide, chloroform, diethylether, iodomethane, perfluoro-1,3-dimethyl-cyclohexanone, and *o*-xylene. Several semivolatile organic compounds were found at concentrations exceeding the minimum detection limits. Those constituents found at the highest concentrations included acetophenone, 3-*tert*-butylphenol, dioctylphthalate, *n,n*-diethyl formamide, diethylene glycol monoethyl ether, *n,n*-dimethyl acetamide, dioctyl phthalate, dipropylene glycol methyl ether, 4-ethylmorpholine, 2-hydroxybenzothiazole, 4-hydroxy-4-methyl-2-pentanone, 1-methyl-2-pyrrolidinone, methylthiobenzothiazole, 2-phenyl-2-propanol, and tetramethylthiourea.

Notably, 72% of the organic carbon in the shuttle condensate samples could be accounted for, but only 21% could be accounted for in the *Mir-19* and *Mir-20* samples.

Samples from the *Mir-21* mission were returned in two batches, one from each of 2 shuttle missions (Pierre et al. 1997). Some volatile organic compounds were present in some samples but not in others, a finding that might reflect the delay between receipt of the two sample batches. Acetone and carbon disulfide were found in all humidity condensate samples. Tetrahydrofuran was found in the *Mir-21* samples but not in any samples from the earlier *Mir* missions. The semivolatile organic compounds benzothiazole, 2-hydroxybenzothiazole, 4-ethylmorpholine, 1-methyl-2-pyrrolidinone, 2-phenyl-2-propanol, 3-*tert*-butylphenol, *n,n*-diethyl formamide, tetramethylthiourea, and tetramethylurea were found in all humidity condensate samples. Other compounds (e.g., caprolactam, diethylformamide, *n,n*-dimethylacetamide, and *n,n*-dimethylformamide) were found in some samples but not in others. Methanol, 1-propanol, 2-propanol, 2-methyl propanol, and 1-methoxy-2-propanol were the alcohols found in all samples. High concentrations of ethylene glycol were traced to a coolant leak in the thermal control system. Formaldehyde was found in significant concentrations (32.5 μg/L). The TOC accountability was only 4-7%, although it was 32% and 67% in the two samples that had high ethylene glycol content.

A discussion of the inorganic contaminants found in humidity condensates from *Mir* are presented later in this chapter, in conjunction with analyses of regenerated water.

Environmental Contributors to Spacecraft Humidity Condensate

Hardware Off-Gassing

Most nonmetallic materials continuously release trace amounts of a wide variety of gases, either through breakdown and subsequent volatilization of the original material or via gradual escape of gases that are trapped in the materials during their manufacture or cleaning. Off-gassed products are of great concern in closed habitats, such as submarines and spacecraft. Compounds off-gassed by articles in spacecraft interiors have included aliphatic hydrocarbons, alcohols, aldehydes, chlorinated hydrocarbons, siloxanes, and carbon monoxide. Even given the constraints on the types of materials that can be flown, gaseous products from the Shuttle and flight articles stored in its lockers could accumulate to harmful concentrations during long missions if they were not removed by the scrubber components of the air revitalization system (Coleman and James 1994).

The types and amounts of chemicals that could be off-gassed from a payload obviously depend on the nature of that payload, and variations make exact predictions difficult. To assess the effect of a "typical" off-gas mixture on recycled water, an ersatz mixture was created and added to the humidity condensate in the WRT. The composition of this ersatz test mixture was based on analyses of humidity condensate from *Spacelab* and shuttle missions (Holder et al. 1995).

Animal Wastes

Life sciences experiments, especially those that involve animals, constitute another source of wastewater contamination. Another mixture of chemicals (an ersatz solution) representing various contaminants found in animal wastes was created on the basis of results from *Spacelab* animal experiments missions (Carter et al. 1995). This mixture (Table 2-5) was added to the waste stream of the Marshall WRT, and the same mixture was used during the 60-d closed-chamber trial of the LMLSTP.

Mechanical Leaks

Recycling potable water from a variety of complex sources is expected to be difficult, and leaks from the water reclamation hardware are likely. Moreover, the potential effects of microgravity on component processes, such as air-fluid separators, vapor compression distillation, catalytic oxidation, and urine pretreatment with highly corrosive

TABLE 2-5 Ersatz Animal Condensate Composition

Compound	Concentration, μg/L
Acetaldehyde	300
Acetone	10,100
Ammonium	590,000
Benzoic acid	850
Bis-2-ethylhexyl phthlate	70
Ethylene glycol	14,000
Nickel	600
Phenol	50
Phosphate	17,000
2-Propanol	11,500
Protein	6,080
Urea	1,170

During the ground-based tests, first in the water recovery tests at Marshall then at Johnson, the animal condensate ersatz solution was metered into the waste bus 24 hr/d at 0.33 lb/hr to simulate condensate input from the research animal holding facility on the space station. It is expected that about 7.92 lb/d would be generated as animal condensate.

Source: Carter et al. (1995).

chemicals, have not been completely identified. Nor are potential system leaks confined to the water system itself. As mentioned above, at least two incidents have taken place on *Mir* in which ethylene glycol has leaked from the thermal control system. It is not surprising that the humidity condensate collected after one of those leaks (1.8 L of 37% ethylene glycol) contained large amounts of ethylene glycol (149.2, 152.8, and 76.8 mg/L) (Pierre et al. 1996).

Microbial Metabolites

Microorganisms will certainly be present aboard any spacecraft that includes humans and animals. Evidence from *Skylab* and the space-shuttle missions indicates that the microbial ecosystem aboard spacecraft undergoes substantial increases in number and perhaps diversity over time (Pierson 1994). Many environmental bacteria can produce volatile metabolites that could end up in the humidity condensate. As an example, bags used to store urine samples for a life sciences experiment on a 1994 shuttle mission were found to give off strong odors. A postflight assessment revealed that the combination of noxious chemicals that had been leaching through the bags could well have come from microbial growth in the urine. Volatile microbial metabolites found in urine storage bags include C8-alkane, C9-alkane, bis(methylthio)methane, carbonyl sulfide, dimethylsulfide, dimethyltrisulfide, ethanol, ethylene glycol, propanone, tetrachloroethene, 1,1,1-trichloroethane, and toluene (J.T. James, Johnson Space Center, personal communication, Jan. 10, 1994).

Several high-efficiency particulate filters in the ECLSS are expected to remove most airborne microbial or fungal particulate matter. However, colonization of the filters and degradation by-products of the colonies might well be carried to the humidity condensate if they cannot be removed effectively by trace contaminant control in the air revitalization system.

Another potential source of microbial metabolites is in the multifiltration resin beds, which contain large amounts of organic carbon. Because large quantities of wastewater will pass through them, these beds may provide a matrix—and a rich nutrient source—for microbial growth and potentially for biofilm formation.

Payload Chemicals

Another source of contaminants in the water system during long space missions is the chemicals carried as part of payload experiments. Toxic chemicals are used frequently in materials-processing and fluid-behavior experiments, in biologic studies that require tissue fixation, and in other types of experiments. Although toxicologists review all flight experiments in detail well before launch, assign hazard levels to component chemicals, and recommend levels of containment, the possibility of escape always exists, especially when experimental equipment is being transferred from one container to another or is being heated. Even a chemical judged to be "acceptable" from a toxicologic standpoint can accumulate in the water system if it escapes.

Utility Chemicals

Volatile chemicals used to disinfect surfaces constitute another source of contamination for the humidity condensate. For example, the high alcohol content in the shuttle humidity condensate presumably reflects the use of alcohol-impregnated "wet wipes" for cleaning interior surfaces (Pierre et al. 1996). A biocide will be used on the ISS to clean surfaces, eating utensils, and windows. The current disinfectant of choice for ISS surfaces is a 2400 parts per million (ppm) Barquat 4250Z solution. This agent, a mixture of *N*-alkyl (C12-C18)-*N*,*N*-dimethyl-*N*-benzyl ammonium chloride and *N*-alkyl (C12-C14)-*N*,*N*-dimethyl-*N*-ethylbenzyl ammonium chloride, has been found to be effective against *Pseudomonos aeruginosa, Staphylococcus aureus,* and *Salmonella choleraesuis.* Other utility chemicals to be considered include glues and adhesives, the off-gas products from which eventually will appear in the water system.

Wash Water and Other Waste Streams

Hygiene Water

Detergents

A major contributor of complex chemicals to the waste stream is hygiene water, which will contain soap, residue from shaving, and chemicals used for oral hygiene. Ingredients of the detergent chosen for ISS personal hygiene and hand washing are listed in Table 2-6.

Chemicals from Personal Hygiene Products

Other personal hygiene items to be used on the ISS, such as shaving cream, toothpaste, and deodorant or antiperspirant, have yet to be standardized in the space-station program. However, some information on use of these products is available from ground-based test beds. During Stage 7 of the WRT, subjects used commercially available shaving cream and toothpaste during a 50-d water-recycling period. No contaminant build-up was noted from the use of these products (Carter and Bagdigian 1993). No build-up was found from the use of similar products, along with a commercially available antiperspirant during the 60-d LMLSTP.

CHeCS Wastes

Another waste stream input to the water processor is the effluent from the CHeCS. This system will be used to provide several routine monitoring procedures over the course of long flights. An ersatz mixture of these compounds was prepared and added to the waste stream of the Marshall WRT, and the same mixture was used during a 60-d closed-chamber trial of the LMLSTP (Meyers et al. 1997).

TABLE 2-6 ISS Soap Ingredients

Ingredient	Percent
Sodium *n*-coconut acid-*n*-methyl taurate (24% active)	98.65[a]
Formaldehyde (formalin 37%)	0.10
Lecipur 95f (soybean lecithin)	0.50
Luviquat FC-500 (polyquaternium)	0.75

[a]CAS No. 104639, RTECS No. GG6500000.
Source: Holder et al. (1995).

Chemicals Formed in the Water Treatment System

Other sources of contaminants in closed-water systems, such as that planned for the ISS, are potential reactions within the system itself. One example noted earlier is the formation of oxidation products from the combination of urine with Oxone and sulfuric acid. Another would be the appearance of chemicals in the treatment train itself that form over long periods of operation, such as fouling in the resin beds. Trimethylamine detected during Stage 10 of the WRT was thought to arise from the reaction of highly basic anions with oxidation chemicals passing through the multifiltration beds or perhaps from large amounts of air being present in the influent to the multifiltration beds (Carter 1997). The bed itself, being a high-carbon source, could form a matrix for biofilm formation. Microorganisms present in the distribution line also could act as a source of microbial metabolites. Biofilms and nonadherent microbial contamination were found in the heat exchangers after a 60-d closed-chamber study in the LMLSTP (D.W. Koenig, Krug Life Sciences, personal communication, April 1997). Even though these organisms were nonpathogenic, they still can produce significant amounts of metabolites. Moreover, the use of biocides in the water system does not preclude the evolution of biocide-resistant species.

System Failures and Incomplete Processing of Influent

Incomplete processing of influent streams can result not only from contaminant overload, but also from mechanical malfunctions of the water system. The design of the integrated system planned for the ISS is extremely complicated, and many components are sensitive to perturbations. The ability to function reliably in microgravity, of course, is critical. Particulate filters are easily clogged. Air-fluid separations are precise processes, and frothing or foaming of detergents will cause poor separations. Poor separation in turn affects the multifiltration beds, causing channeling to adversely affect the flow and the function of the resins, and could even oxidize the resin matrix (e.g., formation of trimethylamine). The vapor compression distillation process depends heavily on motors, complex fluid pumps, and purge pumps to remove noncondensable gases. Frequent pump failures have been experienced during the LMLSTP system tests, which have led to a mixture of coolant and water being processed (data presented at LMLSTP Phase IIA status meeting, NASA/JSC, April 22-24, 1997). Incomplete processing will produce noncondensable gases that will foul the distillation unit and lead to poor quantitative and qualitative recovery of urine distillate. Finally, the multifiltration beds are designed to remove large-molecular-weight organic and ionic components before the water passes to the volatile removal assembly, which catalytically oxidizes the small-molecular-weight organic constituents. Incomplete processing at the multifiltration beds will overload the volatile removal assembly and catalyst with large-molecular-weight organic compounds, eventually producing incomplete products of oxidation.

Chemicals Added to Retard Bacterial Growth

Iodine

Microbial growth and biofilm formation are controlled in the U.S. water system by adding iodine to processed water. About 5 ppm iodine is imparted by a microbial check valve, a polishing ion exchange bed that contains iodinated resin (see Appendix A), which is placed after the volatile removal assembly in the treatment stream. Iodine is added before water is diverted either to the use tank (assuming it meets the criteria of the process control water-quality monitor) or to the potable-water waste-processing stream for reprocessing (if it fails quality criteria). Just as chlorinated by-products can form during disinfection of public water systems, residual iodine can react with a variety of small organic compounds, like phenols, to form iodinated organics while the water is being stored in the ISS. Iodinated organics, like trihalomethanes, can cause adverse health effects (Thorstenson et al. 1987). Even though iodinated organics have not been found in the product water in the Marshall WRT, numerous iodinated organics can form from several organic precursor compounds (Barkley et al. 1992; 1993). High concentrations of phenol, 2-methyl phenol, 3-*tert*-butylphenol, and 2,4-di-*t*-butylphenols were found in the cabin humidity condensates from shuttle and *Spacelab* missions (Muckle et al. 1993); these compounds are potential substrates for the formation of iodinated adducts (Barkley et al. 1992). The adverse health effects from iodine and its by-products have yet to be systematically addressed. Because of the potential for adverse effects from iodine and iodine species, the medical sciences division of the Johnson Space Center has recommended that concentrations of those compounds be reduced at the point of use. Iodine and iodine species have been removed during recent space-shuttle missions by using an iodine removal and mineral injection system or by using activated carbon and ion exchange resins pumped between the shuttle water tank and the galley. Such procedures could become a medical operations requirement for ISS missions.

Silver

The Russian water system uses silver as the residual biocide for processed water. The Russian system was designed to impart silver at concentrations of 0.05-0.5 mg/L of potable water. Water of acceptable quality (i.e., conductivity <150 microsiemens per centimeter) from the humidity condensate processor passes through a conditioning bed, where magnesium, calcium, and other minerals are added for palatability. This proprietary conditioning bed also adds silver to the water. Silver is also added via a silver ionizer to water that is transported from the ground to *Mir*. The treated water is stored in Rodnik tanks and transported by the Russian *Progress* resupply vehicle. The crew uses the water without further processing.

Another source of silver in the water system is the condensing heat exchanger, which collects the humidity condensate. Because the heat exchanger could well support microbial growth, it contains silver components as biocides. Analyses of *Mir* humidity condensate have not shown significant amounts of silver to be present. For example, 3 samples from *Mir-20* had 55-100 µg/L of silver; 1 sample from *Mir-21* had about 500 µg/L; others had no detectable silver. Silver is not analyzed in shuttle humidity condensate samples (Pierre et al. 1996).

REGENERATED WATER

Inorganic Contaminants

A review of federal safe drinking-water standards and health advisories indicates that several inorganic metals and anions and cations can pose risks ranging from simple aesthetic effects (taste, odor, color) to adverse health effects. In the shuttle, potable water is generated electrolytically and concentrations of metal ions are extremely low. This is not the case with the ISS design. There are concerns that the water reclamation process and wastewater treatment processes will result in corrosion and leaching of metal components during storage in potable-water distribution systems. In addition, iodine added to the water as a disinfectant could exacerbate corrosion. The ISS water-quality requirements table (see Appendix A, Table A-1) lists several inorganic compounds and the maximum concentrations to protect crew members from experiencing any potential toxic effects. The multifiltration beds in the *Mir* condensate processor assembly and the Unibed assembly proposed for the U.S. water processor systems, which are made up of combinations of several strong and weak anion and cation exchange resins, were intended to remove inorganic species. Analysis of wastewater samples, and analysis of humidity condensates from several shuttle missions (Muckle et al. 1993; Straub et al. 1995), shuttle-*Mir* missions (Pierre et al. 1996, 1997), and various WRT samples (Carter and Bagdigian 1993) for inorganic contaminants clearly show the presence of various metals and anions and cations. A brief summary of findings from these studies is given below.

Shuttle Humidity Condensates

Samples of humidity condensates from the space shuttle, not processed for crew use, were collected from STS-45, STS-47 (Muckle et al. 1993), and STS-68 (Straub et al. 1995) and analyzed for the presence of metals, anions, and cations. Zinc was found in substantial concentrations in both the cabin (10-20 mg/L) and the *Spacelab* condensates (40.5 mg/L in STS-50, USML-1) (Muckle et al. 1993). Other metals found in the cabin condensates were silicon, nickel, and lithium. It was proposed that the shuttle humidity condensate heat exchangers were the source for these metals. The concentrations differed with the mission and with the age and condition of the heat exchanger. For example, the amounts of nickel in STS-45 (11th mission) humidity condensates were significantly greater than were those in STS-47 (second mission) samples. Silicon was found to come from the water separator filter. Similarly, substantial amounts of zinc were found in condensates from

STS-68 (17-28.5 mg/L). The STS-68 condensates had nickel, copper, lithium, and manganese at low concentrations (< 0.2 mg/L each) but at higher concentrations than reported from previous missions. The anions present in significant concentrations were chloride (0.9-1.8 mg/L) and nitrite (0.45-0.72 mg/L). Ammonium was the major cation in the shuttle condensate at 11-18 mg/L.

Mir Humidity Condensates and *Mir* Reclaimed Water

As a part of water-processing systems development, and to help determine the quality of the raw source water from missions of long duration, humidity condensate samples and samples reclaimed from these condensates using the *Mir* condensate processors were collected from several shuttle-*Mir* missions, starting in March 1995 (named as *Mir-18, Mir-19, Mir-20, Mir-21,* etc.). The analysis of *Mir-20* humidity condensate samples are presented in Table 2-7. A composite summary of inorganic ion concentrations in the humidity condensate samples collected from all the shuttle-*Mir* missions (*Mir-18* to *Mir-25*) is presented in Table 2-8.

TABLE 2-7 Components of Humidity Condensate from *Mir-20*

	Concentration, µg/L	
Component	Minimum	Maximum
Ammonium	4,930	61,800
Calcium	730	2,490
Chloride	320	1,120
Lithium	510	640
Magnesium	340	1,150
Nitrate	470	2,100
Potassium	450	570
Silver	55	100
Sodium	1,250	9,380
Sulfate	240	1,010

Source: Pierre et al. (1996).

TABLE 2-8 Inorganic Compounds with Significant Concentrations in *Mir* Missions

Compound	Concentration, μg/L			
	Minimum	Maximum	Mean	Frequency[a]
Recycled Water				
Barium	ND	2,440	189.7	18/22
Chloride	ND	13,000	3,640	20/23
Fluoride	ND	3,680	206	12/23
Iron	9.3	265	64.4	22/22
Manganese	ND	51.7	9.2	21/22
Nickel	5.6	157	31.3	22/22
Potassium	ND	2,770	621	16/25
Silver	8.4	674	113.6	28/28
Sodium	ND	5,970	1,560	24/25
Sulfate	ND	27,600	3,690	14/23
Humidity Condensate Samples				
Ammonia (as N)	ND	48,000	24,960	26/28
Barium	ND	13,920	635.7	21/23
Calcium	ND	6,110.0	952.8	21/28
Chloride	ND	16,800	1,750	23/28
Magnesium	ND	6,570.0	614	23/28
Nickel	10.2	14,700.0	1,400	23/23
Nitrate as NO_3-N	ND	8,170.0	456.8	7/28
Potassium	ND	20,400	1,800	15/28
Sodium	ND	123,400	8,280	27/28
Sulfate	ND	114,000	6,270	17/28

[a]Number of times detected/total number of humidity condensate samples analyzed from *Mir 18-25* missions.

ND, not detected.

Source: Pierre et al. (1996, 1997, 1999).

Detailed inorganic analysis was not carried out on the reclaimed water from *Mir-18, Mir-19,* and *Mir-20* due to limited sample volume. Chloride, nitrate, sulfate, sodium, ammonium magnesium, and calcium were found. A detailed analysis of condensates and reclaimed water was carried out during *Mir-21*. The results on the inorganic constituents measured in the *Mir-21* humidity condensate samples indicate that most of these compounds were found at concentrations of less than 5 mg/L, except ammonia (as nitrogen), which was detected in one sample at 15.3 mg/L and another at 26 mg/L (as ammonium). Significant concentrations of barium, iron, copper, nickel, and zinc also were found. The zinc concentrations reached a maximum of 5.3 mg/L in one sample. Zinc was detected at concentrations of up to 28.5 mg/L in shuttle condensate (Straub et al. 1995). A composite summary of ionic species identified in the *Mir-18* to *Mir-25* reclaimed water samples shows the presence of chloride, fluoride, sulfate, calcium, silver, nickel, barium manganese, potassium, and sodium (Pierre et al. 1996, 1997, 1999). Those results are included in Table 2-8.

LMLSTP 60-d Product Water

Results of the inorganic analysis on the reclaimed water from the Johnson Space Center 60-d closed-chamber study (LMLSTP Phase IIA) show most of the inorganics well below NASA limits. Only copper, iron, manganese, nickel, and lead were found at significant concentrations, and nickel, copper, manganese and lead exceeded NASA limits (maximum 265, 2170, 71, and 54 μg/L, respectively) (M.E.Homan, J.R. Schultz and R.L. Sauer, NASA/JSC, personal communication, 1999). A cursory examination of shuttle, *Mir*, and ground-based test samples indicates a significant variation in the amount of inorganic constituents in humidity condensate and wastewater. In general, the concentrations in the reclaimed water indicate that the various ion exchange resins included in the processing train (multifiltration bed or Unibed), removed these inorganic constituents to maintain concentrations below the NASA water-quality specifications. Disinfectant iodine concentrations and those for products such as iodides are not included in the tables. The presence of silver in the product waters of Russian *Mir* missions results from the use of silver as a disinfectant in reclaimed water and in water transported via the shuttle.

Nevertheless, the concern that some inorganic metal ions will accumulate in the water is warranted because of the possibility of corrosion over time—especially in the presence of iodine.

Organic Contaminants

Humidity Condensate in *Mir*

Analysis of regenerated hot water collected from point-of-use ports on *Mir-18* and *Mir-19* indicated that the water processor could produce water from condensate that met joint U.S.-Russian specifications for water quality. Nevertheless, TOC, turbidity, and phenol in these samples exceeded the water-quality specifications for the ISS (Pierre et al. 1996). Acetone, bromodichloromethane, chloroform, 1,4-dichlorobenzene, *meta-* and *para*-xylene, and dichlorofluoromethane were the volatile organic compounds found. Formaldehyde was found in samples from only one of the two missions. Table 2-9 lists the organic compounds in *Mir*-regenerated water and their frequency of occurrence.

The organic carbon balance of the regenerated hot water of the two samples from *Mir-18* was 0.47% and 4.58%. From three *Mir-19* samples, it was only 3.3%, 4.0%, and 6.2%.

Two samples of processed water were collected from the galley cold-water ports on *Mir-21* (Pierre et al. 1997). Of the organic compounds detected in these samples, carbon disulfide was the only volatile organic compound found in all samples tested; in contrast, acetone, methylene chloride, and trichloroethene were found in one sample but not in the other. Several semivolatile organic compounds found in processed water from previous *Mir* missions were also found in samples analyzed from this mission.

The recovery of TOC in these two samples again was poor—0.94% for one sample and 12.19% for the other. Explanations offered by the project investigators for the poor recoveries include low amounts of TOC in the product-water samples, small sample volumes, and relatively high detection limits for many components. These factors could combine to produce artifacts in sum calculations for recovery estimates by not accounting for numerous organic species that could be present just below detection limits.

LMLSTP 60-d Product Water

The 60-d Johnson Space Center chamber study, in which four humans used regenerated air and water, was intended to provide a typi-

TABLE 2-9 Organic Compounds in the Processed Water Samples from a Representative *Mir* Mission

Compound	Concentration, μg/L		Frequency[a]
	Minimum	Maximum	
Benzothiazole	1.6	110.7	6
Bis-2-ethylhexyl phthalate	0.4	27.9	6
Di-*n*-butyl phthalate	0.1	1.4	6
Diethyl phthalate	0.1	3.4	6
2-Hydroxybenzothiazole	1.1	6.8	5
2-Mercaptobenzothiazole	8.8	41.8	5
2-Methylthiobenzothiazole	2.2	48.5	5
Toluene	0.8	1.9	5
Urethane	0.4	1.9	5
Acetone	8.89	49.6	4
N-Butylbenzenesulfonamide	0.5	2.7	4
Carbon disulfide	5.2	9.6	4
Decamethylcyclopentasiloxane	0.8	5.6	4
Indole	0.1	0.3	4
Phenol	0.9	1.2	4
N-Phenyl-2-naphthylamine	0.2	11.7	4
1-Tetradecanol	1.8	6.2	4
2,2,4-Trimethyl-1,3-pentanediol diisobutyrate	0.1	0.1	4
Acetate	0.131	1.49	3
Benzyl alcohol	0.9	1.9	3
2,4-Di-*t*-butylphenol	1	1.3	3
Octamethylcyclotetrasiloxane	0.9	1	3
Trichloroethene	1.8	2.3	3
Acetophenone	0.1	41	2
Ethanol	160	2,447	2
Formaldehyde	22.4	63.4	2
2-Phenylphenol	0.2	3.2	2
Phenyl sulfone	0.1	75.1	2
Tetramethylthiourea	0.3	24.6	2
Tris-2-chloroethyl phosphate	0.9	5.8	2
Acetaldehyde	50.85	50.85	1
Benzoic acid	4.6	4.6	1
Butylated hydroxyanisole (BHA)	3.5	3.5	1
3-*tert*-Butylphenol	113.4	113.4	1
Caprolactam	27.4	27.4	1
1,4-Diacetylbenzene	5.6	5.6	1

TABLE 2-9 *(Continued)*

Compound	Concentration, µg/L Minimum	Maximum	Frequency[a]
N,N-Dibutylformamide	8.8	8.8	1
3,5-Di-*t*-butyl-4-hydroxybenzaldehyde	0.1	0.1	1
2,6-Di-*t*-butyl-4-methylphenol	15.8	15.8	1
N,N-Diethylformamide	30.9	30.9	1
N,N-Diethyl-*m*-toluamide	0.5	0.5	1
Diethylene glycol monoethyl ether	25	25	1
N,N-Dimethylacetamide	43.9	43.9	1
N,N-Dimethylbenzylamine	2.1	2.1	1
Dimethylcarbamyl chloride	3.2	3.2	1
Dodecamethylcyclohexasiloxane	1.4	1.4	1
2-Ethoxyethanol	7.7	7.7	1
Ethylene glycol	45,530	45,530	1
2-Ethyl-1-hexanol	1.5	1.5	1
4-Ethylmorpholine	31.8	31.8	1
2-Heptanone	0.4	0.4	1
4-Hydroxy-4-methyl-2-pentanone	11.4	11.4	1
Isophorone	6.3	6.3	1
Methanol	489	489	1
3-Methyl-2-cyclohexen-1-one	6.1	6.1	1
Methylene chloride	4.5	4.5	1
Methyl-4-hydroxybenzoate	5.1	5.1	1
2-Methylpyrazine	6.1	6.1	1
1-Methyl-2-pyrrolidinone	11.4	11.4	1
Methyl sulfone	0.6	0.6	1
sec-Phenethyl alcohol	8.4	8.4	1
2-Phenyl-2-propanol	30.6	30.6	1
Phthalide	2.8	2.8	1
1-α-Terpineol	24.5	24.5	1
Tetramethylsuccinonitrile	31.7	31.7	1
Tetramethylurea	6	6	1
Triethyl phosphate	4.6	4.6	1
Triethylamine	11.8	11.8	1
Urea	1.13	1.13	1
Total organic carbon	5,150	23,700	

[a]Frequency of detection in six samples.
Source: Pierre et al. (1997).

cal demonstration of ISS technology with integrated air and water systems. Because of this, evaluations of the regenerated water contaminants in the product water will provide a better definition of the technology performance (risk mitigation) and a more accurate description of the water contaminants. The compounds identified and the number of occurrences during the 60-d test are presented in Table 2-10. The main purpose of the sorting is to rank the compounds for toxicity evaluation and to assess the processing system to take measures to eliminate contaminants of concern or to control their presence if the source is known. Highly sophisticated onboard monitoring hardware for full characterization will not be available, because of logistical and funding considerations, so the philosophy of the program will be to limit potentially toxic substances entering the system in the first place if the processor can not remove them effectively. The results indicate that the TOC range exceeded NASA's requirement of 0.5 mg/L 16 times, and that values were above 1.0 mg/L five of those times. Acetone and toluene are the only two volatile organic compounds seen of the 57 targeted compounds. Acetone was found four times, but toluene was found in 34 of 68 samples. Among the low-molecular-weight polar hydrophilic compounds that are difficult to remove, formaldehyde was found 30 times; the concentrations, however, were well below the EPA health advisory limit of 1000 μg/L. Other organics—methanol, 1-propanol, 2-propanol, ethanol, and 1,2 propanediol—either were not found or were found infrequently. The results indicate efficient functioning of the volatile removal assembly.

Of particular interest is that compounds that were found often—such as iodoform, diiodomethane, di-*n*-butyl phthalate, ethylhexyl phthalate, and toluene—that are potentially of concern to health were found at low concentrations. The potential for the formation of halomethane-type compounds with iodine as the disinfectant is clearly indicated. In general, the lessons learned in the 60-d chamber test in the areas of component functioning and the efficiency of contaminant removal systems to potable-water quality will help NASA finetune the design and build a robust and dependable system for long space missions.

Stage-10 WRT

The Marshall Space Flight Center WRT Stage-10 was conducted for 128 d in a recipient mode of operation, in which reclaimed water was

TABLE 2-10 Organic Contaminants in Processed Waters During the 60-d Johnson Space Center Chamber Study

Compound	Concentration, μg/L		Frequency[a]
	Minimum	Maximum	
Iodoform	1.6	4.8	40
Methyl sulfone	0.6	54.5	38
Di-*n*-butyl phthalate	0.1	2.3	36
Toluene	1.01	9.53	34
2-Ethyl-1-hexanol	0.4	12.4	33
Formaldehyde	2	13.8	30
Benzaldehyde	0.1	0.8	24
1-Methyl-2-pyrrolidinone	0.3	7.4	21
Benzyl alcohol	0.8	7	21
Diiodomethane	0.4	1.8	19
Diisopropyl adipate	0.4	0.9	17
Dodecamethylcyclohexasiloxane	0.5	1.7	13
Acetate	0.14	3.97	11
Bis-2-ethylhexyl phthalate	0.1	1.7	11
4-Hydroxy-4-methyl-2-pentanone	0.6	3.4	10
Methanol	101	233	8
1-Formylpiperidine	0.1	0.6	7
Squalene	0.9	3.9	6
Decamethylcyclopentasiloxane	0.1	0.2	5
Diethyl phthalate	0.2	0.4	5
Lactate	0.18	1.1	4
Oxalate	0.23	0.41	4
sec-Phenethyl alcohol	0.1	0.2	4
Acetone	4.64	6.3	3
Acetophenone	0.1	0.3	3
Benzothiazole	0.1	0.7	3
Pentacosane	0.3	1.2	3
Phenol	0.7	1	3
2-(2-Butoxyethoxy)ethanol	0.4	0.8	2
2-Butoxythanol	3.1	3.2	2
2-Ethylhexanoic acid	1.7	2.1	2
Monomethyl phthalate	4.8	4.8	2
N,N-Dimethylbenzylamine	0.6	0.6	2
Octamethylcyclotetrasiloxane	0.4	0.6	2
Octanoic acid	2.6	2.9	2
Tris-2-chloroethyl phosphate	0.9	1.5	2
Benzylbutyl phthalate	3.6	3.6	1

TABLE 2-10 *(Continued)*

	Concentration, µg/L		
Compound	Minimum	Maximum	Frequency[a]
Bis-2-ethylhexyl adipate	0.8	0.8	1
n-Butyl palmitate	5.3	5.3	1
Butyl stearate	10.7	10.7	1
1,4-Diacetylbenzene	0.3	0.3	1
Neomenthol	0.2	0.2	1
2-Phenyl-2-propanol	0.5	0.5	1
2-Propanol	154	154	1
Tetramethylsuccinonitrile	0.4	0.4	1
Tributyl phosphate	0.5	0.5	1
Urea	302	302	1
Total organic carbon	87	1,850	

[a]Frequency of detection in 68 samples.

Source: J. Schultz and colleagues, Wyle Laboratories, personal communication, Sept. 1997.

returned to test subjects for use. The water recovery system processed pretreated urine flush water; an ersatz sample of CHeCS waste; ersatz animal condensate; humidity condensate; ersatz equipment off-gas; ersatz fuel cell water; and wet shave, personal hygiene, and oral hygiene water. The results in general indicated no accumulation of contaminants in the product water (Carter 1997). When they become available, detailed analysis data on organic species will be reported.

The issues involved with developing appropriate water-processing hardware and treatment technologies are complex. Because of the unique nature of the input water to the ISS water processor system, NASA has placed strong emphasis on characterizing the nature of potential contaminants and on assessing variations in the influent streams as thoroughly as possible. Several efforts have been made to predict the chemical makeup of source waters to the ISS water processor system.

Ground-based test beds, such as the WRT and the LMLSTP, can simulate only some aspects of ISS input water. Humidity condensate samples have been collected from a few space-shuttle and *Mir* missions, but the constituents of these spacecraft samples have not been compared with those of the ground-based test beds. A composite compilation of wastewater data from various sources has been attempted (Carter 1998).

Despite substantial variability in TOC accountability between shuttle and *Mir* missions, and the very poor accountability from the *Mir* product water (partly due to detection limits issues), the results gathered so far from ISS simulated wastewater streams and in-flight raw and processed water samples indicate that the organic compounds that are present are vastly different from the list of target compounds developed by EPA for public drinking water.

MONITORING WATER CONTAMINANTS

Conductivity is the only process control monitoring currently done aboard *Mir* in its condensate processor system. During the ISS early assembly phase, to monitor the water quality, the Russian segment service module will accommodate U.S.-provided water-quality monitoring hardware for in-flight off-line monitoring with provisions for archiving water samples for ground-based analysis. The U.S. program proposes several strategies for water monitoring after the ISS is assembled.

Source-Water Monitoring

As noted by the NRC (1992), the drivers for designing and operating any potable-water processor will depend on the organic and inorganic constituents of the raw source water. Adequate specifications for ensuring the quality of the reclaimed or recycled water do not exist outside of NASA, and thus attention must be directed to the definition of source water. Because the content of source waters aboard spacecraft cannot be predicted reliably from ground-based experiments, the NRC recommended that actual humidity condensate and other raw waters be collected from short- and long-term space flights. The waters should be thoroughly analyzed, and their variability should be clarified. This task also is important in terms of defining potential health hazards from the use of water aboard spacecraft.

Described below are monitoring strategies derived from continuing tests of isolated and integrated water-recovery systems conducted at the Marshall Space Flight Center (Bagdigian et al. 1991; Carter et al. 1992; Homan et al. 1994; Holder et al. 1995), analyses of humidity condensate collected from the space shuttle (Muckle et al. 1993; Straub et al. 1995) and from *Mir* (Pierre et al. 1996), and analyses of samples from a human-rated life-support chamber (the LMLSTP) (J Schultz, Wyle Laboratories, personal communication, Sept. 1997).

In-Line Monitoring and Process Control

Conductivity is the only aspect of water quality that is monitored routinely aboard *Mir* and planned for the early phase 2 of the ISS assembly in the Russian service module. In the current proposed configuration of the ISS, an in-line process control water-quality monitor (PCWQM) will assess conductivity, iodine, pH, and TOC. (Product-water TOC should be less than 0.5 mg/L.) If the parameters are within acceptable limits, the effluent is transferred to the potable-water subsystem; if not, the water is shunted upstream of the multifiltration subsystem for reprocessing (see Appendix A, Figure A-3). Current plans for monitoring ISS water quality after completion of the ISS (plans that are subject to revision) are outlined in Appendix A, Table A-3.

Off-Line Monitoring

The ISS will have one tank, and if the water limit fails the criteria by the on-line monitor it will be reprocessed.

At present, only conductivity and other surrogate quality measures can be used for routine in-line water-quality monitoring. (A water-quality monitor that can measure conductivity, TOC, and iodine—all pass/fail criteria for potability—has been tested successfully in Stage 10 of the Marshall WRT (Carter 1997).) A comprehensive evaluation (complete characterization) of the product water might be needed after every recycling to ensure that the technologies can effectively remove compounds that are potentially hazardous to the health of crew members. Even though organic and inorganic constituents will be removed by the series of granular activated-carbon and ion-exchange beds, and low-molecular-weight compounds will undergo catalytic oxidation at the volatile removal assembly, trace contaminants could produce iodinated organic compounds through reactions with residual iodine in the water. Because such disinfection by-products have neither been chemically characterized nor assessed for potential toxicity, the precursor materials that could contribute to the formation of the by-products should be well understood.

Complete characterizations of product water, particularly its organic constituents, will require off-line analyses, which in turn require that hardware and analytical methods be developed for use in space. Progress toward that goal was provided by the 60-d test of the LMLSTP in a human-rated regenerative life support chamber at Johnson Space Center (Meyers et al. 1997). This test bed was used to develop a comprehensive strategic tier for monitoring aspects of processed-water quality off-line, such as tests for process verification, and selective and complete characterization (Table 2-11).

Even though several forms of water-quality monitoring hardware initially were proposed by the CHeCS for providing detailed in-flight analyses, programmatic and logistical considerations have been restricted to only a few for off-line monitoring (Table 2-12).

TABLE 2-11 Water Analyses for Phase IIA of the Lunar-Mars Life Support Test Project

Water Verification	Potable Characterization[a]	Complete Characterization[b]
Conductivity	Alcohols	Anions
Iodine, iodide	Amines	Cations
TOC	Carboxylates	Color
pH	Formaldehyde	Glycols
	Nonvolatile organic compounds	Mercury
	Organic acids	Metals
	Semivolatile organic compounds	Turbidity
	Volatile organic compounds	Urea

[a]Includes water verification.

[b]Includes water verification and potable characterization.

TOC, total organic carbon

Source: Master Protocol for the Participation of Human Test Subjects in Phase IIA of the Early Human Testing Initiative, Crew and Thermal Systems Division, Johnson Space Center, Oct. 16, 1996.

TABLE 2-12 ISS Water Monitoring Capabilities by Crew Health Care Systems

Instrument	Parameter	Purpose
TOC analyzer	TOC, TIC, TC, pH, conductivity	Assess organic load in reclaimed water
Sampler and archiver	NA	Sample from water systems for inflight and ground analyses
Spectrophotometer	Color, turbidity, iodine species; UV/Vis spectra; wet chemistry	Assess general water quality and U.S. biocide concentrations and effectiveness

The first two items will support monitoring during the early phases of ISS when the Russian service module and life support modules will process water.

NA, not applicable; TC, total carbon; TIC, total inorganic carbon; TOC, total organic carbon; UV/Vis, ultraviolet-visible

Sources: International Space Station Medical Operations Requirements Document (ISS MORD), Document Number SSP 50260, Baseline, January 1998; and Crew Health Care System (CHeCs) GFE Specification, International Space Station, Revision Basic, June 1999, NASA Johnson Space Center (Draft) Document Number SSP 50470.

SUMMARY

The joint U.S.-Russian space program was expanded to include collecting samples of condensate and reclaimed cold and hot water during flights, beginning with *Mir-18* in 1995 and continuing through *Mir-25* in March 1998. Samples were collected with hardware supplied by the U.S., and they were analyzed by both NASA (at the Johnson Space Center) and the Russian Space Agency (at the Institute of Biomedical Problems). Results have shown that the reclaimed water meets all Russian water-quality specifications (except for TOC) and most NASA standards (except for some halogenated hydrocarbons). However, only 5% TOC in samples collected from *Mir* could be identified with available analytical techniques.

To date, only humidity condensate has been used to regenerate potable water during flight in *Mir*. Information on water recovered from *Mir* urine processors under microgravity conditions is not available, and the only microgravity opportunities to test the technologic aspects of water reclamation are *Mir* missions. Conceptual and functional differences between the U.S. and Russian approaches, as well as logistical problems in exchanging information about technology, are challenging. In addition, the materials used in the *Mir* modules might not be the same as those planned for the ISS, and *Mir* information is 7-10 years old. Results from a ground-based, human-rated chamber facility at Johnson Space Center, which had integrated systems for air revitalization and water recovery, provided useful supplements for microgravity tests from a systems operations point of view.

Several issues remain to be resolved for the ISS program, particularly differences between the Russian and U.S. systems in the quantity of water to be produced and in the quality standards for that water. However, in the past 2 years, as a result of periodic technical interchange meetings between Russian and U.S. scientists, several issues concerning water quality, monitoring, and quantity are being resolved. The Russian water processor that will be launched during phase 2 (on the Russian service module) will have a slightly different design from that on *Mir*, and water quality will be judged solely on the basis of conductivity. Planned modifications to the Russian urine-processing system include the distillation process, the design and function of the heat pump, optimization of the air flow rate, and increases in capacity and reliability. Weekly *Mir* system status updates provided to NASA on the performance of both the condensate processors (SRV-K and the SRV-U) have provided a mechanism to assess confidence in the systems' performance.

Long-term concerns include the build-up of chemical contaminants that are difficult to remove during water processing and the possibility of progressive microbial colonization of the multifiltration resin systems and distribution systems. Concerns also have been raised about the possibility of the generation of iodine-resistant microorganisms and their metabolic products in the water system. Finally, the planned in-flight water-quality monitoring hardware is limited and perhaps far from mature in terms of its microgravity compatibility testing and flight qualifications.

Combinations of extensive ground-based testing completed so far involving human crews and risk mitigation experiments aboard the shuttle and *Mir* have provided additional information on processes and processors with which to provide water that poses minimal long-term health risks to crew members. Comprehensive comparisons of those chemicals found in regenerated water from all forms of tests should form the basis for a hazard assessment database for specific chemical constituents in product water. This will be a complex task, as a variety of issues must be considered, including differences in the exposure scenarios between ground-based and in-flight studies (i.e, microgravity, use of certain drugs); differences in exposure among in-flight and ground-based studies (i.e., duration, cabin materials); delays between sampling and analysis of in-flight water; and better technology for more recent tests.

REFERENCES

Bagdigian, R.M., M.S. Traweek, and G.K. Griffith. 1991. Phase III Integrated Water Recovery Testing at MSFC: Partially Closed Hygiene and Open Potable Loop Results and Lessons Learned. SAE Technical Paper Series no. 911375, 21st International Conference on Environmental Systems, San Francisco, CA, July 1991.

Barkley, R., C. Hurst, A. Dunham, J. Silverstein, and G.M. Brion. 1992. Generation of Iodine Disinfection By-products (IDPs) in a Water Recycle System. SAE Technical Paper Series no. 921362, 22nd International Conference on Environmental Systems, Seattle, WA, July 1992.

Barkley, R., A. Dunham, J. Silverstein, and C. Hurst. 1993. Iodine Disinfection By-products (IDPs) Generated in Water from Selected Organic Precursor Compounds. SAE Technical Paper Series no. 922097, 23rd International Conference on Environmental Systems, Colorado Springs, CO, July 1993.

Carter, D.L. 1997. Phase III Integrated Water Recovery Testing at MSFC: International Space Station Recipient Mode Test Results and Lessons Learned. SAE Technical Paper Series no. 972375, 27th International Conference on Environmental Systems, Lake Tahoe, NV, July 14-17, 1997.

Carter, D.L. 1998. Waste Water Characterization for the ISS Water Processor. SAE Technical Paper Series no. 981616, 28th International Conference on Environmental Systems, Danvers, MA, July 13-16, 1998.

Carter, D.L., and R.M. Bagdigian. 1993. Phase III Integrated Water Recovery Testing at MSFC: Single-loop Test Results and Lessons Learned. SAE Technical Paper Series no. 932048, 23rd International Conference on Environmental Systems, Colorado Springs, CO, July 1993.

Carter, D.L., H. Cole, M. Habercom, and G. Griffith. 1992. Determination of Organic Carbon and Ionic Accountability of Various Waste and Product Waters Derived from ECLSS Water Recovery Tests and Spacelab Humidity Condensate. SAE Technical Paper Series no. 921313, 22nd International Conference on Environmental Systems, Seattle, WA, July 1992.

Carter, D.L., D.W. Holder, and C.F. Hutchens. 1995. International Space Station Environmental Control and Life Support System, Phase III Water Recovery Test Stage 9, Final Report. NASA Tech Memo 108498. NASA Marshall Space Flight Center, Huntsville, AL.

Cole, H., M. Habercom, S. Crenshaw, S. Johnson, S. Manuel, W. Martindale, G. Whitman, and M. Traweek. 1991. The Characterization of Organic Contaminants During the Development of the Space Station Water Reclamation and Management System. SAE Technical Paper Series no. 911376. 21st International Conference on Environmental Systems, San Francisco, CA, July 1991.

Coleman, M.E., and J.T. James. 1994. Airborne toxic hazards. Pp. 141-156 in Space Physiology and Medicine, 3rd Ed., A.E. Nicogossian, S.L. Pool, and C.L. Huntoon, eds. Philadelphia: Lea & Febiger.

EPA (U.S. Environmental Protection Agency). 1996. Drinking Water Regulations and Health Advisories. EPA 822-B-96-002. U.S. Environmental Protection Agency, Washington, DC.

Holder, D.W., D.L. Carter, and C.F. Hutchens. 1995. Phase III Integrated Water Recovery Testing at MSFC: International Space Station Configuration Test Results and Lessons Learned. SAE Technical Paper Series no. 951586, 25th International Conference on Environmental Systems, San Diego, CA, July 1995.

Homan, M.H., P.D. Mudgett, J.R. Schultz, and R.L. Sauer. 1994. GC/MS and CE Methods for the Analysis of Trace Organic Acids in Reclaimed Water Supplies. SAE Technical Paper Series no. 941392, 24th International Conference on Environmental Systems, Friedrichshafen, Germany, July 1994.

Homan, M.E., J.R. Schultz, and R.L. Sauer. 1997. Water Analysis Results from Phase II of the NASA Early Human Testing Initiative 30-day Closed Chamber Test. SAE Technical Paper Series no. 972555, 27th International Conference on Environmental Systems, Lake Tahoe, NV, July 1997.

James, J.T., T.F. Limero, H.J. Leano, J.F. Boyd, and P.A. Covington. 1994. Volatile organic contaminants found in the habitable environment of the Space shuttle: STS-26 to STS-55. Aviat. Space Environ. Med. 65:851-857.

Meyers, K.E., D.G. Staat, and F.D. Smith. 1997. Lunar-Mars Life Support Test Project Phase II-A: A Crew's 60-day Experience. SAE Technical Paper Series no. 972340, 27th International Conference on Environmental Systems, Lake Tahoe, NV, July 14-17, 1997.

Mitchell, K.L., R.M. Bagdigian, R.L. Carraquillo, D.L. Carter, G.D. Franks, D.W. Holder, C.F. Hutchens, K.Y. Ogle, J.L. Perry, and C.D. Ray. 1994. Technical Assessment of Mir-1 Life Support Hardware for the International Space Station. NASA TM 108441. NASA Marshall Space Flight Center, Huntsville, AL.

Muckle, S.J., J.R. Schultz, and R.L. Sauer. 1993. Characterization of Spacecraft Humidity Condensate. SAE Technical Paper Series no. 932176, 23rd International Conference on Environmental Systems, Colorado Springs, CO, July 1993.

NRC (National Research Council). 1986. Space Station Water Quality. Conference held at NASA Johnson Space Center, Houston, TX, July 1-2, 1986. Report JSC 32024, March 1987.

NRC (National Research Council). 1989. Chemical and Water Quality and Monitoring Requirements of Reclaimed Water for Space Station Freedom. Meeting sponsored by NASA Johnson Space Center, Medical Sciences Division, Houston, TX, Nov. 27-29, 1989.

NRC (National Reserach Council). 1992. Chemical Water Quality and Monitoring Requirements of Reclaimed Water for Space Station Freedom. Washington, DC: National Academy Press.

PHS (U.S. Public Health Service). 1962. Public Health Service Drinking Water Standards, Revised Ed. U.S. Department of Health Education and Welfare, U.S. Public Health Service, Rockville, MD.

Pierre, L.M., J.R. Schultz, S.M. Johnson, R.L. Sauer, Y.E. Sinyak, V.M. Skuratov, and N.N. Protasov. 1996. Collection and Chemical Analysis of Reclaimed Water and Condensate from Mir Space Station. SAE Technical Paper Series no. 961569, 26th International Conference on Environmental Systems, Monterey, CA, July 1996.

Pierre, L.M., J.R. Schultz, S.M. Johnson, R.L. Sauer, Y.E. Sinyak, V.M. Skuratov, and N.N. Pratasov. 1997. Chemical Analysis of Potable Water and Humidity Condensate Collected During the Mir-21 Mission. SAE Technical Paper Series no. 97-ES-224, 27th International Conference on Environmental Systems, Lake Tahoe, NV, July 1997.

Pierre, L.M., J.R. Schultz, R.L. Sauer, Y.E. Sinyak, V.M. Skuratov, N.N. Pratasov, and L.S. Bobe. 1999. Chemical Analysis of Potable Water and Humidity Condensate: Phase One Final Results and Lessons Learned. SAE Technical Paper Series no. 1999-01-2028, 29th International Conference on Environmental Systems, Denver, CO, July 12-15, 1999.

Pierson, D.L. 1994. Microbiology. Pp. 157-166 in Space Physiology and Medicine, 3rd Ed., A.E. Nicogossian, C.L. Huntoon, and S.L. Pool, eds. Philadelphia: Lea & Febiger.

Putcha, L., S.L. Pool, and N.M. Cintron. 1994. Pharmacology. Pp. 435-446 in

Space Physiology and Medicine, 3rd Ed., A.E. Nicogossian, C.L. Huntoon, and S.L. Pool, eds. Philadelphia: Lea & Febiger.

Samsonov, N.M., L.S. Bobe, V.M. Novikov, N.S. Farafonov, G.K. Abramov, B.Y. Pinsky, E.I. Grigorov, E.N. Zaitsev, N.N. Protasov, V.V. Komolov, A.I. Grigoriev, and Y.E. Sinyak. 1997. Updated Systems for Water Recovery from Humidity Condensate and Urine for the International Space Station. SAE Technical Paper Series no. 972559, 27th International Conference on Environmental Systems, Lake Tahoe, NV, July 14-17, 1997.

Straub, J.E., J.R. Schultz, W.F. Michalek, and R.L. Sauer. 1995. Further Characterization and Multifiltration Treatment of Shuttle Humidity Condensate. SAE Technical Paper Series no. 951685, 25th International Conference on Environmental Systems, San Diego, CA, July 1995.

SSP (Space Station Program). 1995. International Space Station Flight Crew Integration Standard NASA STD 3000T. SSP 50005 Rev. B. Aug. Johnson Space Center, Houston, TX.

SSP (Space Station Program). 1996. Segment Specification for the US On-Orbit. SSP 41162E, p. 273. July. Johnson Space Center, Houston, TX.

SSP (Space Station Program). 1998. International Space Station Medical Operations Requirements Document (ISS MORD). SSP 50260. Jan. Johnson Space Center, Houston, TX.

SSP (Space Station Program). 1999. Crew Health Care System (CheCs) GFE Specification, International Space Station, Revision Basic. SSP 50470. July. Johnson Space Center, Houston, TX.

Thorstenson, Y.R., D.S. Janik, and R.L. Sauer. 1987. Medical Effects of Iodine Disinfection Products in Spacecraft Water. SAE Technical Paper Series no. 871490, 17th International Conference on Environmental Systems, July 1987.

Verostko, C.E. 1986. Space Station Water Sources. Paper presented at the Space Station Water Quality Conference, Lyndon B. Johnson Space Center, Houston, TX, July 1-2, 1986.

Verostko, C.E., K. Pickering, F. Smith, N. Packham, J. Lewis, G. Stonesifer, D. Staat, and M. Rosenbaum. 1997. Performance of the Water Recovery System During Phase II of the Lunar-Mars Life Support Test Project. SAE Technical Paper Series no. 972417, 27th International Conference on Environmental Systems, Lake Tahoe, NV, July 1997.

Winkler, H.E., C.E. Verostko, and G.F. Dehner. 1983. Urine Pretreatment for Water Processing Systems. SAE Technical Paper Series no. 831113.

3

Sources and Types of Data for Establishing Spacecraft Water Exposure Guidelines

IN this chapter, the Subcommittee on Spacecraft Water Exposure Guidelines describes the sources and types of data that should be used for establishing spacecraft water exposure guidelines (SWEGs). This information is similar to that described in the National Research Council's (NRC 1992) *Guidelines for Developing Spacecraft Maximum Allowable Concentrations for Space Station Contaminants* (SMACs), but with important differences. First, the major route of exposure considered in SMACs is inhalation, whereas the major expected route of exposure considered in SWEGs is oral, via ingestion of drinking water and food prepared in potable water. Dermal absorption and inhalation could be secondary routes of exposure. Second, the SWEGs should incorporate advances in the understanding of human physiology and metabolism in microgravity that have occurred since the SMACs were published. Third, the duration of the exposures of interest is different for contaminants in drinking water, because exposure can be avoided if necessary by not drinking the water from the potable drinking-water supply of the spacecraft (emergency supplies could be made available for short periods). It is much more difficult to avoid contact with the ambient air in a spacecraft.

As noted in Chapter 2, contamination of the water used for consumption and personal hygiene could result either from the release of toxic substance into the atmosphere with subsequent absorption into water (e.g., into water condensate collected for reclamation) or because of a malfunction of the water purification system.

Several types of information are evaluated to develop risk-based SWEGs for water, including data on the physical and chemical properties of the substance; human clinical and epidemiologic studies; animal toxicity data from studies, in which exposure ranges from acute to chronic, to identify toxic and carcinogenic end points; in vitro toxicity studies; and mechanistic studies. Data from oral exposure studies are the most relevant for deriving SWEGs, but dermal absorption and inhalation studies should also be evaluated because exposure from those routes occurs when there are volatile constituents in the water or when water is used for hygiene purposes. In addition, those types of data can be used to predict oral toxicity, in the absence of relevant oral data, when toxicokinetic and metabolic data are available to predict disposition and toxicity after oral administration.

CHEMICAL AND PHYSICAL CHARACTERISTICS OF A TOXICANT

The chemical and physical characteristics of a chemical influence its absorption, distribution, metabolism, and excretion from the body. When they are expressed quantitatively, the effect is called the pharmacokinetics (or toxicokinetics) of the compound. For example, molecular size, stability in stomach acid, water solubility, and overall lipophilicity of ingested substances strongly influence their ability to be absorbed and their subsequent pharmacokinetics. More information is needed to determine whether microgravity or other factors alter the physiologic disposition of toxic substances based on their chemical and physical properties.

Ideally, SWEGs should be established for all compounds that might be found in water onboard spacecraft. As a practical matter, it would be difficult and costly to develop SWEGs for the more than 400 chemical species identified to date in the *Mir* space-station water system. So, the National Aeronautics and Space Administration (NASA) must identify the compounds that cause the most concern and rank them accordingly for SWEG development. Those criteria are discussed in Chapter 5.

HUMAN STUDIES

Human toxicity data frequently are obtained from epidemiologic studies of low-level, long-term environmental and industrial exposures as well as short-term exposures, usually to large quantities of toxicants after accidents. These data sometimes provide a basis for estimating a dose-response relationship.

However, epidemiologic studies have several limitations. Most studies involve retrospective analyses that provide only a small amount of data on past chemical exposures (Checkoway et al. 1989). For example, in occupational studies, exposures are typically estimated from records based on the number of years of employment in a given place or from available records on the work environment or personal samples (Rinsky 1989). Prospective studies, which involve assessment of exposure while the cohort is followed, can provide more reliable information on exposure because the sampling scheme can be devised as part of the research plan rather than relying on available data collected for other purposes, such as assessing compliance with exposure regulations (Smith 1987). However, these studies also have limitations because some outcomes of exposure are not manifested for months or even years after exposure.

Epidemiologic studies also vary in the accuracy and precision of the health outcome measured. Some outcomes must rely on information collected for other purposes, such as death certificates, which can show an error rate of 20-40% in the United States (Percy et al. 1981); early preclinical testing; pathology reports; or early preclinical markers of pathology. Despite these limitations, if the populations studied are large enough, have had substantial exposure, and have had a sufficient interval between exposure and study to allow for the expression of disease, epidemiologic studies offer the advantage of providing human data. Epidemiologic studies often can provide data that assist in establishing a permissible concentration for human exposure (e.g., Threshold Limit Values of the American Conference of Governmental Industrial Hygienists).

Epidemiologic outcomes often are reported in terms of relative risk, a ratio of the rate of outcome of disease or disability in the exposed population to that in the nonexposed population. Care must be taken in using this type of information because relative risk is not a measurement of risk. For example, the relative risk for a rare disease can be the same as that for a common disease and lead to substantially less total risk. To determine an acceptable exposure level, information on relative risk should be analyzed to identify the relationship of relative risk and risk of morbidity or mortality. The risk that is acceptable is a matter of policy and could vary significantly depending on which population is exposed.

ANIMAL STUDIES

Most risk assessments use data derived from laboratory animal studies combined with human clinical and epidemiologic data, when available. To assess either the acute or the chronic toxicity of a water contaminant, emphasis should be placed on human data, provided that it is sufficiently reliable. Using data directly from humans obviates the need to estimate relative sensitivities of humans and animals to the toxic effects of a substance. However, exposure-response data for humans often are not available, and extrapolation from animal data is necessary. The most useful animal studies are those in which the exposure occurs by relevant routes of administration (oral for drinking-water contaminants; dermal absorption and inhalation for contaminants found in water used for purposes of hygiene) and in which the duration of exposure approximates human exposure times. Confidence in extrapolation from animals to humans is increased if at least two nonhuman species have been examined and if the physiologic disposition (including metabolic pathway), target organs, and toxic effects in animals parallel the effects expected in humans based on available information. NASA will use recently developed techniques for combining uncertainty factors to increase confidence in extrapolations from animal data (Chapter 4). For 1- and 10-day (d) SWEGs, data from acute toxicity tests in animals should be used; for 100- and 1000-d SWEGs, the reference should be subchronic, chronic, or lifetime studies in animals.

When considering the use of published studies on the toxicity of a chemical in establishing a SWEG, attention must be paid to species and, in some cases, to strain differences as they relate to the applicability of assessing health risks in humans. There might be quantitative or qualitative differences in xenobiotic-metabolizing activity between laboratory animals and humans. There also could be mechanistic considerations to suggest that one species is more predictive than is another of responses in humans.

Exposure via oral feeding or gavage can provide a scientifically sound and defensible basis for estimating effects in humans and for predicting the concentrations at which those effects occur. Studies that use water as the vehicle would be most useful. Repeated exposures to a test substance are useful in the identification of homeostatic adaptations or repair and recovery that could occur over time. It is important to note the doses and to give special attention to evaluation of extremes in dosage. Dosage regimen—gavage versus feeding or drinking water, repeated versus single—and the possibility of additional exposure from inhalation should be considered. Oral dosing experiments permit the testing of hypotheses about the mechanism of the toxic action of the pollutant. Ideally, sufficient data should be collected to establish a no-observed-adverse-effect level (NOAEL) or to establish an accurate estimate of a benchmark dose (BMD). The BMD has a specified low level of excess health risk, generally in the range of 1%-10% that can be estimated from data with little or no extrapolation outside the experimental dose range.

In many studies the substance is administered by gavage on an acute, subchronic, or chronic basis in a vehicle, often water or an oil, in the form of a bolus dose directly into the stomach within a brief period. A major difficulty in evaluating gavage studies is that blood concentrations and attendant effects are induced that might not be observed if the administration were spread out over several smaller doses, as would be expected with the normal pattern for water consumption. The metabolism and pharmacokinetics associated with the single, high doses might be different from what would be observed if repeated, lower doses were used. The resulting absorption might be influenced by the vehicle, or the vehicle itself could have an adverse effect on the animal. The relative absorption from water and food must be considered in evaluating animal studies and in estimating human exposure to contaminants in drinking water and water-reconstituted food.

Subchronic studies that assess regular treatment with the chemical over 90 days (typically) permit examination of cumulative effects while also permitting time for repair mechanisms or physiologic adaptation. Toxic effects on specific organs can be evaluated as a function of dose and time of administration in a given species, the dose at the target organ, and the likelihood of accumulation of effects over time. Ideally, toxic effects should be related to several factors, including total dose, number of treatments with the test substance, and frequency of administration.

Toxic effects for a specific substance might be different in animals exposed to the contaminant by repeated exposure to low doses over an extended period than they will be in animals exposed to a single, high dose. For example, acute exposure to benzene can cause depression of the central nervous system, bu repeated exposure can result in leukemia. Thus, the most sensitive end point and data for establishing a 1000-d SWEG can be completely different from sensitive end point and data appropriate for establishing a 1-d SWEG.

Chronic studies cover most of the lifetime of an animal, and can range from a relatively short duration to lifetime exposure. They allow examination of effects at low doses and can be used to detect accumulated toxic effects or repair mechanisms that activate after an extended period of exposure. Carcinogenic and noncarcinogenic end points alike can be evaluated. Special attention must be directed to how well results can be extrapolated to humans. Although most effects observed in animals are also seen in humans, there are some examples that do not extrapolate to humans, such as the nephropathy-related appearance of α_{2u}-globulin in rat kidney attendant to exposure to a series of compounds (many of which are components of motor fuels). Another example is the frequent occurrence of primary liver tumor in C3H mice observed in many bioassays despite the fact that primary liver tumor is rarely observed in humans and for the most part is not known to be caused by similar exposures. Data from experiments in which the maximum tolerated dose is used must be interpreted with care because human exposures do not mimic that paradigm of carcinogenesis bioassays. In the interpretation of chronic-exposure bioassays, mechanisms of carcinogenesis, species specificity, examination of threshold levels, and the operation of initiation versus promotion must be integrated to provide the best judgment regarding potential human toxicity.

The same considerations and concerns about subchronic studies apply to chronic studies. Additional attention should be directed to the influence of naturally occurring conditions in test animals in altering the disposition and the metabolism of the substance during the course of the study. Potential physiologic alterations with age must be addressed.

IN VITRO TOXICITY STUDIES

Important information can be obtained from studies that investigate adverse effects of chemicals on cellular or subcellular systems in vitro. Systems in which toxicity data have been collected include isolated organ systems (e.g., isolated perfused livers and lungs), single-cell organisms, cells isolated from specific organs of multicellular organisms and maintained under defined conditions (e.g., isolated hepatocytes and bone-marrow colony-forming units), functional units derived from whole cells (e.g., organized subcellular particles), breakdown products of cellular disruption (e.g., microsomes and submitochondrial particles), isolated or reconstituted enzyme systems, and specific macromolecules (e.g., proteins and nucleic acids).

In vitro studies can be used to elucidate the toxic effects of chemicals and to provide information on their mechanism of action. Typically, in vitro systems are used on the assumption that effects observed are reasonable models for humans. However, for NASA's purposes, the additional caveat that they should reflect the response of humans in space must be added. Thus, in vitro studies must be interpreted with caution, because the effects of microgravity on cellular systems must be considered.

ADVANCES IN HEALTH EFFECTS ASSESSMENT

In setting SWEGs, all observed toxic effects should be considered: mortality, morbidity, functional impairment, reproductive toxicity, developmental toxicity, genotoxicity, carcinogenicity, neurotoxicity, immunotoxicity, hepatotoxicity, respiratory toxicity, and in vitro toxicity. Developmental toxicity data are included in the analysis for comprehensiveness, even though pregnant astronauts are barred from spaceflight. The various toxicity effects are reviewed extensively in the SMAC guidelines (NRC 1992) and are not repeated here. However, there have been several advances that are applicable to SWEGs in the areas of neurobehavioral toxicology, reproductive toxicology, and mutagenesis. Those advances are discussed below.

Neurobehavioral Effects

Several reports emphasize the importance of neurobehavioral testing in assessing the effects of chemical pollutants, including water contaminants, on the central nervous system (Moser and MacPhail 1992; MacPhail and Tilson 1995). For the most part, such assessment provides noninvasive measures of sensory-motor performance (speed, accuracy, fine discrimination). Some water contaminants pose a potential health hazard because of their ability to alter nervous system function and impair performance of complex tasks, as shown in studies on occupational exposure (Anger 1990).

Extrapolation from industrial thresholds and other related exposure standards for such determinations is limited by the use of tests based on gross toxic effects under conditions of discontinuous exposure that seldom consider the more subtle effects revealed by performance impairment. Moreover, data on the effects of water contaminants on human performance are rarely available, and the interactive influences of stressful, physically demanding spaceflight environments are undetermined. There is a range of human factors and conditions that can be expected to hamper neurobehavioral adaptation and enhance vulnerability to the toxic effects of environmental pollutants (NRC 1987). The factors include confined space, lack of privacy, weightlessness requirements for readjustment of motor and perceptual skills, disorientation, and space sickness. They occur under conditions of isolation, demanding workloads, and the ever-present danger of being away from Earth.

There is now broad acceptance that functional observational batteries and motor activity tests in animals provide a reliable screen for neurotoxic compounds. These assessments include evaluating clinical signs and measuring motor activity, schedule-controlled behavior, and morphologic change in the nervous system (Sette 1989; Holson et al. 1990; Sette and MacPhail 1992; Moser et al. 1995). Motor activity patterns measured automatically provide a continuous, noninvasive assessment of a water contaminant's effects on a stable performance baseline over an extended interval. In addition, schedule-controlled behavior based on the programming of performance antecedents and consequences can provide specific measures of learning and memory function (McMillan and Leander 1976; MacPhail 1994) as well as sensory thresholds and reaction times (Brady et al. 1979).

Of particular relevance for detecting and evaluating water contaminants is the increasing recognition of taste aversion conditioning (Garcia et al. 1961; Domjan 1980) as a relatively simple, sensitive, and reliable measure of neurobehavioral effects. When animals are exposed to water contaminated with toxins after they have consumed water flavored with a novel substance (saccharin), they subsequently avoid consuming saccharin-flavored water. Such conditioned flavor aversions have been convincingly demonstrated with a variety of toxic water contaminants, including methylmercury (Levine 1978), cadmium (MacPhail 1982a), cobalt (Wellman et al. 1984), trialkyltins (Leander and Gau 1980; MacPhail 1982b), thallium (Nachman and Hartley 1975; Peele et al. 1986), copper (Nachman and Hartley 1975), arsenic (Rzoska 1953), and lead (Dantzer 1980; Leander and Gau 1980). The results of these studies also suggest that taste aversion procedures might provide a novel and selective paradigm for assessing the interactive effects of water contaminants and purification agents.

The effectiveness of animal tests for predicting the human response to metal chelators, for example, has been demonstrated in reported studies with BAL (British antilewisite or 2,3-dimercaptopropanol) and DMSA (2,3-dimercaptosuccinic acid), a water-soluble BAL analogue (Peele et al. 1987). The finding that BAL was significantly more potent than DMSA in conditioning flavor aversions reflects both the reported differences in the toxicity of the two chelators (Aposhian 1983) and the high and low incidence of clinical side effects (fever, nausea, hypertension) with BAL and DMSA, respectively (Klaassen 1980). There is also a good correspondence between the human response to abusable drugs and preclinical animal tests results with such substances (Brady 1991). Despite the predictive promise of laboratory assessment models, however, more study is needed to evaluate the functional relationship between the results obtained with animals in neurobehavioral toxicity tests and effects in humans.

Morphologic changes in the nervous system can be expected to reveal more serious neurobiologic effects of exposure to environmental contaminants, including water pollutants (e.g., Katz et al. 1981; Johnson and Richardson 1983; Spencer and Schaumburg 1999). Such invasive assessments, however, require both in situ perfusion and labor-intensive contemporary tissue preparation for microscopic examination.

Reproductive Effects

The interpretation and application of reproductive toxicity data are difficult if the intent is to apply results to the determination of SWEGs. First, there are two sexes to consider, and each has a different set of reproductive targets for potential toxicants. Targets include the various cells involved in oogenesis and spermatogenesis, the viability and functional capacity of the mature ovum and sperm, fertilization, implantation, and gestation. The interaction of adenohypophyseal-hypothalamic axis and peripheral reproductive organs can be affected as well. Whereas some of the effects can be grossly and readily observed, others are more subtle and might not become apparent until some time after children of the astronaut are born.

Although there are several tests to evaluate reproductive capacity for each sex, exhaustive studies will not have been performed for most compounds of interest in establishing SWEGs. Nevertheless, for the male, data might be available on morphologic and functional parameters of the testis, epididymis, or accessory sex glands; measurements of semen; or measurements of the concentrations of various hormones. For females, a variety of measurements might be available on effects on morphology or function; the oviduct; or hormonal activity, including effects on the hypothalamus, pituitary, or hormones, such as gonadotropin, chorionic gonadotropins, estrogen, or progesterone. Other measures might include implantation, teratological changes, fetotoxicity, and postnatal determination of impairments. The U.S. Environmental Protection Agency has established guidelines for reproductive toxicity assessment (EPA 1996), and there are several publications that provide guidance on evaluating and interpreting reproductive toxicity end points for human health risk assessment (e.g., ILSI 1999; NRC in press).

It will be necessary to evaluate the dose and regimen of exposure, the route of administration, and the relevance of the test result to the human condition for each toxic effect related to exposure before data can be applied to the calculation of a SWEG.

Mutagenesis

Available mutagenesis data on a variety of cell types exposed to water contaminants in vivo or in vitro provide information that is useful for determining whether genetic risks should be included in setting concentration guidelines on contaminants in space-station drinking water. Three main types of mutations serve as good indices for mutagenic potential: single-gene mutation, resulting from a change in the molecular structure of DNA, which can result from substitution or loss or gain of a single base pair (genetic polymorphism); loss or rearrangement of large segments of DNA caused by strand breakage (chromosome aberrations); and a change in the amount of DNA in a cell (aneuploidy). In the first two types, DNA is the target molecule; for aneuploidy induction, the proteins in spindle microtubules and centromeres are the most likely targets.

Mutations in germ cells cause heritable abnormalities in offspring. Because these cells in mammals of both sexes are sequestered from circulating blood (more so in the female than in the male), the probability seems small that low concentrations of an ingested chemical will induce mutations in germ cells. In contrast, somatic cells have ready access to blood, as evidenced by the highly mutagenic compounds used for tumor chemotherapy. There are now data showing that a mutation produced in a single somatic cell could result in cancer. Thus, in the space-station environment, mutation that leads to carcinogenesis is the primary concern.

Evidence that mutation is integral to all aspects of carcinogenesis, from initiation through promotion and progression, is now overwhelming (Loeb 1996; Bishop 1997); in fact, mutation is the driving force in the loss of control over cell division and movement. Most tumors are clones; they are derived from mutation in a single cell (Barrett 1995). However, many subsequent mutations, including those that involve mutator genes (Minnick and Kunkel 1996), participate in the complex, multistep, long-term carcinogenic process that leads to a clinically recognizable tumor.

The altered gene products primarily involved in carcinogenesis are generally the proteins that control cell division, growth, and movement, thereby promoting clonal expansion and metastasis. Some of them are transcription factors, tumor suppressors, mitotic check point controllers, chromosome structure stabilizers, DNA damage recognition proteins and repair enzymes, apoptotic inhibitors, chaperonins, and proteases that break down intercellular adhesion proteins. An example is a mutated tumor suppressor gene, *p53*, found in 40-50% of common tumors, such as those of the lung, colon, and breast (Perera 1997). Its normal function is to regulate other genes so that growth of a cell with damaged DNA is inhibited until repair is completed or, in the absence of repair, signals production of proteases that kill the cell (apoptosis).

New test systems for mutagenic and carcinogenic chemicals have been developed that considerably improve sensitivity; some of the more prominent are listed here:

- The comet assay is a single-cell gel electrophoresis technique for detecting DNA damage (Plappert et al. 1997).
- Immunomarkers and multifluorescent probes for specific genes and for all 24 human chromosomes can reveal small increments in mutation frequency. For example, aneuploidy for individual chromosomes can be readily detected in sperm of mammals, including humans (Robbins et al. 1997), and translocation chromosomes can be unequivocally identified in dividing cells (Yang et al. 1997).
- Transgenic animals heterozygous for a proto-oncogene (+/-), such as the *p53* gene, permit unequivocal in vivo identification of carcinogenic chemicals that cause cancer by mutating one *p53* gene; nonmutagenic, noncarcinogenic chemicals also can be identified (Tennant et al. 1996).
- DNA site-specific techniques use oligodeoxynucleotides that contain a single chemical-derived DNA adduct that helps identify postsynthetic changes, such as deletions and base substitutions in a gene (Shibutani and Grollman 1997).

The high specificity of polymorphic tumor mutants, and the sensitivity of new tests for identifying them, emphasize the relevance of mutagenesis to chemical carcinogenesis. Detection of low levels of mutagenicity combined with positive biomarker data (Perera 1997) provides strong evidence for mutagenic potency. Data obtained with new methods combined with results from earlier studies enhance confidence in using mutagenicity data in determining SWEGs.

Much attention and research are being devoted to establishing carcinogenic risks on the basis of mutagenic data (see Dellarco and Jacobson-Kram 1996; Gold and Zeiger 1996), and there is some optimism that a formulation can be constructed for the recognized vital relationship of the two processes. In the meantime, establishing human risk levels attending germ and somatic cell mutations induced by exposure to various contaminants must rely on the "weight of evidence" method.

MECHANISTIC STUDIES

Health effect studies are required by regulatory agencies before many substances can be marketed for various purposes in the United States. They fall into the general category of descriptive toxicology. With the maturation of toxicology as a distinct discipline, the investigation of the biology that leads to toxic events has emerged as a significant focus of research. This area of research is called mechanistic toxicology. There is a major effort to apply the results of mechanistic studies to the risk-assessment process. Experiments in mechanistic toxicology might include, but not be limited to, studies of routes of administration, absorption, metabolism, excretion, tissue distribution, formation of biologic reactive intermediates, covalent binding to specific macromolecules, physiologically based pharmacokinetic models, and polymorphisms.

The recent development of investigative tools in molecular biology has opened new ways to evaluate mechanisms of toxicity. It is now possible to study directly the interaction of specific substances with the genome itself, or to study enzymes that modify gene function. The ability to identify and measure a large new class of small proteins, such as the cytokines and related controlling factors, through the use of new methods in immunochemical analysis permits the study of mechanisms by which the toxicity of some substances might be expressed as interfering with the normal functioning of these products. The use of knockout and transgenic animals requires fine judgments to ensure that the data can be extrapolated to normal animals and exposed humans. An excellent example of this application comes from a report on a study of mice in which the gene for the synthesis of CYP2E1 was knocked out. The animals were no longer sensitive to benzene-induced bone-marrow depression, because CYP2E1 is the enzyme responsible for metabolic activation of benzene to metabolites that inhibit bone-marrow function (Valentine et al. 1996).

Classic toxicity studies use normal animals on Earth. In the controlled conditions of the laboratory, attention is paid to their diurnal rhythms and to light and dark cycles, which might not mimic the situation of NASA's chief concern, the astronaut in space. It might be necessary to conduct laboratory studies under conditions that simulate space flight or to develop animal models with features similar to the physiologic state of the astronaut in prolonged spaceflight. Although such models are approximations to the human condition, they might provide more relevant information than could studies on unaltered normal animals. Animals that have flown in space are likely to be more appropriate surrogates for humans. For example, rats flown aboard *Cosmos 1887* showed altered hepatic function (Merrill et al. 1990). They also demonstrated skeletal muscle weakness resulting from muscle fiber atrophy and segmental necrosis. In addition to the microgravity and radiation of space, animals are exposed to launch and reentry gravity forces, noise, and vibration.

Although data from oral exposures to toxic substances are preferred to develop SWEGs for drinking water, data from experiments in which toxicants are administered by other routes are potentially useful. Because species can differ in their responses to toxic substances, the utility of animal data depends in part on the species used. For example, aflatoxin B_1 induces liver tumors in rats, hamsters, and monkeys but not in mice (IARC 1993). The mechanistic basis for the difference is thought to be related to species differences in the expression of a particular form of the enzyme glutathione S-transferase in the liver (Eaton and Gallagher 1994). In the absence of information on target organs and pharmacokinetics in both animals or humans, however, the confidence in an extrapolation from animals to humans can be low. As relevant human data are accumulated they should be incorporated into the SWEGs process.

SUMMARY

In developing SWEGs, several types of data should be evaluated, including the physical and chemical characteristics of the contaminant, human clinical and epidemiologic studies, in vitro toxicity studies, toxicokinetic studies, animal toxicity studies conducted over a range of exposure durations, genotoxicity studies, and carcinogenicity bioassays. All observed toxic effects should be considered, including mortality, morbidity, functional impairment, neurotoxicity, immunotoxicity, reproductive toxicity, developmental toxicity, genotoxicity, and carcinogenicity. For completeness, developmental effects should be considered in the analyses, even though pregnant astronauts are barred from space flight.

Data from oral exposure studies should be used, particularly drinking water and feed studies, in which the duration of exposure approximates human exposure times. Dermal absorption and inhalation studies should also be evaluated, as exposure from those routes can also occur from water.

There are several important determinants for deriving a SWEG, including identifying the most sensitive target organ or body system affected; the nature of the effect on the target tissue; dose-response relationships for the target tissue; the rate of recovery; the nature and severity of the injury; cumulative effects; toxicokinetic data; interactions with other chemicals; and the effects of microgravity.

REFERENCES

Anger, W.K. 1990. Worksight behavioral research. Results, sensitive methods, test batteries and the transition from laboratory data to human health. Neurotoxicology 11(4):627-717.

Aposhian, H.V. 1983. DMSA and DMPS—Water soluble antidotes for heavy metal poisoning. Ann. Rev. Pharmacol. Toxicol. 23:193-215.

Barrett, J.C. 1995. Role of mutagenesis and mitogenesis in carcinogenesis. Pp. 21-32 in Environmental Mutagenesis, D.H. Phillips and S. Venitt, eds. Oxford: BIOS Scientific Publishers.

Bishop, J.M. 1997. Cancer: What should be done? Science 278:995.

Brady, J.V. 1991. Animal models for assessing drugs of abuse. Neurosci. Biobehav. Rev. 15(1):35-43.

Brady, J.V., L.D. Bradford, and R.D. Heinz. 1979. Behavioral assessment of risk-taking and psychophysical functions in the baboon. Neurobehav. Toxicol. 1(Suppl 1):73-84.

Checkoway, H., N. Pearce, and D.J. Crawford-Brown. 1989. Research Methods in Occupational Epidemiology. New York: Oxford University Press. 344 pp.

Dantzer, R. 1980. Conditioned taste aversion as an index of lead toxicity. Pharmacol. Biochem. Behav. 13(1):133-135.

Dellarco, V.L., and D. Jacobson-Kram. 1996. New Approaches for Cancer Identification and Risk Assessment [Editorial Introduction]. Mutat. Res. 365:1-2.

Domjan, M. 1980. Ingestional aversion learning: Unique and general processes. Pp. 275-336 in Advances in the Study of Behavior, Vol. 11, J.S. Rosenblatt, R.A. Hinde, C. Beer, and M. Busnel, eds. New York: Academic Press.

Eaton, D.L., and E.P. Gallagher. 1994. Mechanisms of aflatoxin carcinogenesis. Annu. Rev. Pharmacol. Toxicol. 34:135-172.

EPA (U.S. Environmental Protection Agency). 1996. Guidelines for Reproductive Toxicity Risk Assessment. Fed. Regist. 61(212):56273-56322.

Garcia, J., D.J. Kimeldorf, and E.L. Hunt. 1961. The use of ionizing radiation as a motivating stimulus. Psychol. Rev. 68(6):383-385.

Gold, L.S., and E. Zeiger. 1996. Handbook of Carcinogenic Potency and Genotoxicity Databases. Boca Raton, FL: CRC Press.

Holson, R.R., M.G. Paule, and F.M. Scalzo. 1990. Methods in behavioral toxicology and teratology. Neurotoxicol. Teratol. 12(5):413-567.

IARC (International Agency for Research on Cancer). 1993. IARC Monographs on the Evaluation of Carcinogenic Risks to Humans. Some Naturally Occurring Substances: Food Items and Constituents, Heterocyclic Aromatic Amines and Mycotoxins, Vol. 56. Lyon, France: International Agency for Research on Cancer.

ILSI (International Life Sciences Institute). 1999. An Evaluation and Interpretation of Reproductive Endpoints for Human Health Risk Assessment, G. Daston, and C. Kimmel, eds. Washington, DC: ILSI Press.

Johnson, M.K., and R.J. Richardson. 1983. Biochemical endpoints: Neurotoxic esterase assay. Neurotoxicology 4(2):311-320.

Katz, R., C.N. Tai, R.M. Diener, R.F. McConnell, and D.E. Semonick. 1981. Dichloroacetate, sodium: 3-month oral toxicity studies in rats and dogs. Toxicol. Appl. Pharmacol. 57(2):273-287.

Klaassen, C.D. 1980. Heavy metals and heavy-metal antagonists. Pp. 1615-1637 in The Pharmacological Basis of Therapeutics, 6th Ed., A.G. Gilman, L.S. Goodman, and A. Gilman, eds. New York: Macmillan.

Leander, J.D., and B.A. Gau. 1980. Flavor aversions rapidly produced by inorganic lead and triethyl tin. Neurotoxicology 1:635-642.

Levine, T.E. 1978. Conditioned aversion following ingestion of methylmercury in rats and mice. Behav. Biol. 22(4):489-496.

Loeb, L.A. 1996. Many mutations in cancers. Pp. 329-342 in Genetic Instability in Cancer, Cancer Surveys, Vol. 28, T. Lindahl, ed. Plainview, NY: Cold Spring Harbor Laboratory Press.

MacPhail, R.C. 1982a. Flavor aversions induced by cadmium: Importance of route of exposure [abstract]. Toxicologist 2(1):4.

MacPhail, R.C. 1982b. Studies on the flavor aversions induced by trialkyltin compounds. Neurobehav. Toxicol. Teratol. 4(2):225-230.

MacPhail, R.C. 1994. Behavioral analysis in neurotoxicology. Pp. 7-18 in Neurobehavioral Toxicity: Analysis and Interpretation, B. Weiss, and J.L. O'Donoghue, eds. New York: Raven Press.

MacPhail, R.C., and H.A. Tilson. 1995. Behavioral screening tests: Past, present, and future. Pp. 231-238 in Neurotoxicology: Approaches and Methods, L. Chang, and W. Slikker, Jr., eds. San Diego: Academic Press.

McMillan, D.E., and J.D. Leander. 1976. Effects of drugs on schedule-controlled behavior. Pp. 85-139 in Behavioral Pharmacology, S.D. Glick, and J. Goldfarb, eds. St. Louis, MO: C.V. Mosby.

Merrill, A.H., Jr., M. Hoel, E. Wang, R.E. Mullins, J.L. Hargrove, D.P. Jones, and I.A. Popova. 1990. Altered carbohydrate, lipid, and xenobiotic metabolism by liver from rats flown on Cosmos 1887. FASEB J. 4(1):95-100.

Minnick, D.T., and T.A. Kunkel. 1996. DNA synthesis errors, mutators and cancer. Pp. 3-20 in Genetic Instability in Cancer, Cancer Surveys, Vol. 28, T. Lindahl, ed. Plainview, NY: Cold Spring Harbor Laboratory Press.

Moser, V.C., and R.C. MacPhail. 1992. International validation of a neurobehavioral screening battery: The IPCS/WHO collaborative study. Toxicol. Lett. 64/65:217-223.

Moser, V.C., B.M. Cheek, and R.C. MacPhail. 1995. A multidisciplinary approach to toxicological screening: III. Neurobehavioral toxicity. J. Toxicol. Environ. Health 45(2):173-210.

Nachman, M., and P.L. Hartley. 1975. Role of illness in producing learned taste aversions in rats: A comparison of several rodenticides. J. Comp. Physiol. Psychol. 89(9):1010-1018.

NRC (National Research Council). 1987. A Strategy for Space Biology and Medical Science for the 1980s and 1990s. Washington, DC: National Academy Press.

NRC (National Research Council). 1992. Guidelines for Developing Spacecraft Maximum Allowable Concentrations for Space Station Contaminants. Washington, DC: National Academy Press.

NRC (National Research Council). In press. Evaluating Chemical and Other Agent Exposures for Reproductive and Developmental Toxicity. Washington, DC: National Academy Press.

Peele, D.B., R.C. MacPhail, and J.D. Farmer. 1986. Flavor aversions induced by thallium sulfate: Importance of route of administration. Neurobehav. Toxicol. Teratol. 8(3):273-277.

Peele, D.B., J.D. Farmer, and R.C. MacPhail. 1987. Conditioned flavor aversions: Applications in assessing the efficacy of chelators in the treatment of heavy-metal intoxication. Toxicol. Appl. Pharmacol. 88(3):397-410.

Percy, C., E. Stanek, 3rd, and L. Gloeckler. 1981. Accuracy of cancer death certificates and its effect on cancer mortality statistics. Am. J. Public Health 71(3):242-250.

Perera, F.P. 1997. Environment and cancer: Who are susceptible? Science 278(5340):1068-1073.

Plappert, U.G., B. Stocker, H. Fender, and T.M. Fliedner. 1997. Changes in the repair capacity of blood cells as a biomarker for chronic low-dose exposure to ionizing radiation. Environ. Mol. Mutagen. 30(2):153-160

Rinsky, R.A. 1989. Benzene and leukemia: an epidemiological risk assessment. Environ. Health Perspect. 82:189-191.

Robbins, W.A., M.F. Vine, K.Y. Truong, and R.B. Everson. 1997. Use of fluorescence in situ hybridization (FISH) to assess effects of smoking, caffeine, and alcohol on aneuploidy load in sperm of healthy men. Environ. Mol. Mutagen. 30(2):175-183.

Rzoska, J. 1953. Bait shyness: A study in rat behavior. Br. J. Anim. Behav. 1(4):128-135.

Sette, W.F. 1989. Adoption of new guidelines and data requirements for more extensive neurotoxicity testing under FIFRA. Toxicol. Ind. Health 5(2):181-194.

Sette, W.F., and R.C. MacPhail. 1992. Qualitative and quantitative issues in assessment of neurotoxic effects. Pp. 345-361 in Neurotoxicology, H. Tilson, and C. Mitchell, eds. New York: Raven Press.

Shibutani, S., and A.P. Grollman. 1997. Molecular mechanisms of mutagenesis by aromatic amines and amides. Mutat. Res. 376(1-2):71-78.

Smith, T.J. 1987. Exposure assessment for occupational epidemiology. Am. J. Ind. Med. 12(3):249-268.

Spencer, P.S., and H.H. Schaumburg. 1999. Experimental and Clinical Neurotoxicology, 2nd Ed. New York: Oxford University Press.

Tennant, R.W., J. Spalding, and J.E. French. 1996. Evaluation of transgenic mouse bioassays for identifying carcinogens and noncarcinogens. Mutat. Res. 365(1-3):119-127.

Valentine, J.L., S.S. Lee, M.J. Seaton, B. Asgharian, G. Farris, J.C. Corton, F.J. Gonzalez, and M.A. Medinsky. 1996. Reduction of benzene metabolism and toxicity in mice that lack CYP2E1 expression. Toxicol. Appl. Pharmacol. 141(1):205-213.

Wellman, P.J., P.A. Watkins, J.R. Nation, and D.E. Clark. 1984. Conditioned taste aversion in the adult rat induced by dietary ingestion of cadmium or cobalt. Neurotoxicology 5(2):81-90.

Yang, T.C., K. George, A.S. Johnson, M. Durante, and B.S. Fedorenko. 1997. Biodosimetry results from space flight Mir-18. Radiat. Res. 148(Suppl. 5):S17-S23.

4

Risk Assessment Methods for Determining Spacecraft Water Exposure Guidelines

HUMAN exposure guidelines for toxic substances are established through a multiple-step process called *risk assessment*. The guidelines are set for concentrations that research predicts pose acceptable (usually negligible) risks of adverse health effects to humans under specified conditions of exposure. Quite often, the objective of risk assessment is to establish a daily exposure that is considered safe over a lifetime. For space travel, the anticipated durations are substantially less than a lifetime, but the absolute lifetime risk of adverse health effects is still the focus of the risk assessment. For adverse effects that are transitory and only mildly debilitating, the intent is to ensure that exposure to substances that cause such effects is restricted to amounts that will not impede the normal performance of duties aboard spacecraft.

Although the process of risk assessment uses human data whenever possible, often from epidemiologic studies, it is not aimed at estimating relative risks in the usual epidemiologic sense. Risk assessment frequently involves extrapolation from conditions under which the data are derived by observation to an unobserved or unobservable exposure situation, and it focuses on absolute risk rather than relative risk. More often than not, because of the lack of suitable human data, risk assessment is based on data from experiments with animals. The quality of the data has a major influence on the risk assessment, and meaningful extrapolation of experimental data to an applicable human situation presents significant challenges.

Below is a review of the approaches to conducting risk assessments, as well as the subcommittee's recommended approach to deriving spacecraft water exposure guidelines (SWEGs). A discussion of the exposure conversions and uncertainty factors that should be considered in the calculations is also provided.

HISTORICAL PERSPECTIVE

Risk Assessment for Noncarcinogenic Effects

For toxic effects other than cancer, the practice of risk assessment has been to set acceptable exposure by dividing no-observed-adverse-effect levels (NOAELs) obtained from human studies or animal experiments by a set of uncertainty factors (sometimes called "safety" factors). A NOAEL is the highest experimental dose for which no difference in the occurrence of an adverse effect is observed relative to a control group. The NOAEL-based approach has come to be associated with the presumed existence of threshold doses—doses below which specific toxic effects will not occur, even if exposure continues over a lifetime. The concept of threshold is supported by the observation that many organisms have detoxification mechanisms or repair capacities to compensate for some degree of damage and still maintain normal function (Klaassen and Eaton 1991). Exposure guidance levels that result from reducing NOAELs by uncertainty factors, called *acceptable daily intakes* or ADIs, are presumed to pose zero risk of the toxic effect in question. In many applications, two uncertainty factors of 10 have been thought to be adequate, the first to allow for possible increased sensitivity of humans to the toxic agent compared with experimental animals and the second to account for variations in susceptibility within the human population (Lehman and Fitzhugh 1954).

In experiments for which a NOAEL is not established, only a lowest-observed-adverse-effect level (LOAEL) will be available for risk assessment. A LOAEL generally corresponds to a response in the range of 1-

10%, and an uncertainty factor of 10 often is used to extrapolate from a LOAEL to a NOAEL, although some investigators indicate that a factor of 3-5 would be more appropriate (Abdel-Rahman and Kadry 1995). Ideally, the selection of the uncertainty factor depends on the slope of the dose-response curve.

Barnes and Dourson (1988) identified two additional uncertainty factors that might be needed for deriving references doses (RfDs), which estimate a daily exposure to the human population that is likely to be without an appreciable risk of harm during a lifetime. These additional factors represent uncertainty with respect to exposure duration and to data quality. The size of each of several uncertainty factors is determined by the best judgment of the risk assessor; however, the U.S. Environmental Protection Agency (EPA) has suggested using a maximum of 3000 for the product of four uncertainty factors and a maximum of 10,000 for five uncertainty factors (Dourson 1994). Uncertainty factors involved in the calculation of SWEGs are discussed later in this chapter.

Risk Assessment for Carcinogenic Effects

It has been assumed traditionally that threshold doses do not exist for carcinogenic effects, particularly those considered to result from genotoxicity. For this reason, it has been considered infeasible to establish low exposure limits that correspond to zero risk. Instead, beginning with the pioneering work of Mantel and Bryan (1961), attempts have been made to estimate carcinogenic risks on a precise, quantitative basis, to estimate exposures that produce very low, but nonzero, cancer risks. These efforts have involved fitting mathematical models to experimental data and extrapolating downward to predict risks at doses well below the experimental range.

The mathematical model most frequently used for low-dose extrapolation is a variation of the multistage model of Armitage and Doll (1960), commonly expressed as

$$P(d) = 1 - \exp(-q_0 - q_1 d - q_2 d^2 - \ldots - q_k d^k)$$

$P(d)$ is the probability of developing cancer during a lifetime of exposure at a dose d of a carcinogen, and $q_0, q_1, q_2, \ldots, q_k$ are nonnegative constants that are estimated via regression analysis of cancer data (usually animal data) at *k* or more dose levels. Ideally, *d* is a measure of target tissue dose, but most often in practice *d* is a measure of external exposure.

According to the multistage theory, a malignant cancer cell develops in stages from a single stem cell through a series of biologic events (mutations) that occur in a specific order. Assuming that the rates of transition between two or more stages in the multistage model are linearly related to target tissue dose, the dose-response curve for the multistage model is linear at low doses (Crump et al. 1976). Low-dose linearity is generally assumed for chemical carcinogens that operate through direct interaction with genetic material. When carcinogenesis occurs by other mechanisms, low-dose linearity might not be applicable. Data developed in recent years suggest that some carcinogens, especially those whose mechanisms of action involve cytotoxicity or disruption of hormonal homeostasis, exhibit practical threshold doses below which the risk of cancer is negligible (Page et al. 1997; Hill et al. 1998). However, if there is a nonzero background cancer risk and if the mechanism of the nongenotoxic carcinogen is the same as the background mechanism, then this "additivity" of cancer risk still implies linearity at very low doses for dose-response relationships that are strictly increasing. Hence, linear extrapolation has been widely used in low-dose cancer risk assessment in the absence of clear information to dictate a different course of action (OSTP 1985; EPA 1996a). Risk assessments that deviate from the use of linear extrapolation require considerable data to ensure that the traditional default approach is not applicable.

Uncertainties in the process of establishing acceptable exposures have been handled differently for carcinogenic and noncarcinogenic effects. For example, instead of a factor of 10 for interspecies uncertainty, a factor derived from a power function of body weight often is used for interspecies conversion (EPA 1992). Also, in general, no uncertainty factor is used for human variation in sensitivity to a substance's carcinogenic effects. However, variation in the experimental data is a source of uncertainty that is recognized for carcinogenic effects through the use of statistical confidence limits instead of central estimates.

RECOMMENDED APPROACH TO RISK ASSESSMENT

Exploiting Similarities of Historical Approaches

Gaylor (1983) was among the first to point out the practical similarities between low-cancer-risk dose based on linear extrapolation and those that would result from the reduction of cancer NOAELs by uncertainty factors. In light of Gaylor's observation that the NOAEL for cancer often corresponds to a central estimate of risk of approximately 10^{-2} (1%), then *reducing such a cancer NOAEL by an overall uncertainty factor* of 100, 1000, or 10,000 would result in the same dose that would be obtained by linear extrapolation to a risk level of 10^{-4}, 10^{-5}, or 10^{-6}, respectively (Kodell and Park 1995). Conversely, if the true response rate at the NOAEL for a noncarcinogenic effect is acknowledged to be other than zero, say around 1%, then the *application of linear extrapolation for a noncarcinogenic effect* to estimate a dose with a risk level k orders of magnitude lower would be equivalent to dividing the NOAEL by an uncertainty factor of 10^k. The functional equivalence of the two approaches has been highlighted by Wilson (1997).

Despite the apparent practical similarities between the two opposing approaches to risk assessment, little has been done until recently to unify them. Proponents of low-dose linear extrapolation have questioned the presumption by NOAEL proponents that zero-risk limits (thresholds) can be established based on experimental observations; proponents of the NOAEL-uncertainty factor approach have questioned the presumption by modeling proponents that precise risks can be attached to doses below the observed experimental range. Recently, however, proposals have been advanced for the unification of risk assessment procedures for carcinogenic and noncarcinogenic effects (Purchase and Auton 1995; Crump et al. 1996; Gaylor et al. 1999). There is a movement to place less emphasis on numerical estimates of risk of cancer below the data range, and to place more emphasis on estimation of risk of noncancer effects within the data range. The objective is to combine the best features of the two methods into a unified approach to setting safe exposure for all types of toxic effects.

Exploration and development of refined models for low-dose extrapolation is not discouraged. Rather, as biologic processes are better understood, it is expected that improved mathematical models for risk assessment will evolve. Several promising new approaches are discussed later in this chapter. However, the usual data that are available for risk assessment do not permit precise estimates of risk to be made at doses below the data range. For this reason, the risk assessment methodology presented here emphasizes model fitting within the data range for carcinogenic and noncarcinogenic effects.

Benchmark-Dose Approach to Setting SWEGs

It is important that dose-response data are adequate to establish a NOAEL. However, various authors have documented limitations of the NOAEL as a basis for establishing acceptable exposure (Munro and Krewski 1981; Crump 1984; Kimmel and Gaylor 1988). The determination of the NOAEL is limited by the number and distribution of doses and by sample sizes; using the NOAEL as a basis for setting exposure limits ignores dose-response information. As an alternative to the NOAEL, Crump (1984) proposed use of a *benchmark dose* (BMD)—a statistical lower confidence limit on a dose that is estimated to correspond to a low level of excess risk above background in the range of 1-10% (ED_{01} to ED_{10}; ED_p is an effective dose that yields a response of p). Because of its accounting for experimental variation, the BMD could be lower than the NOAEL and thus could result in lower acceptable exposure limits after it has been reduced by uncertainty factors. Experimental variation, however, is an important source of uncertainty that has been neglected heretofore in risk assessment for noncarcinogenic effects.

The BMD originally was defined as a statistical lower confidence limit on the ED_p, for $0.01 \leq p \leq 0.10$. In addition to the original suggestion of Crump (1984), observations by other investigators also argue for establishing the BMD to correspond to the response range between 1% and 10%. As observed by Gaylor (1992) and Allen et al. (1994a), the incidence of fetal malformation at the NOAEL in typical teratology studies often exceeds 1%. Leisenring and Ryan (1992) argue that the average risk at the NOAEL for quantal data could easily be as much as 10%, depending on the experimental design and the shape of the dose-response curve. In an analysis of 486 developmental toxicity studies, Allen et al. (1994a) conclude that the average NOAEL approximated a lower 95% confidence limit on the ED_{05}. Several investigators recommend the ED_{01} as an anchor point for risk estimates for carcinogenic effects (Van Ryzin 1980; Farmer et al. 1982; Gaylor et al. 1994). The intent is to avoid dependence on particular mathematical models, which is most apparent at doses below the ED_{01} (Krewski and Van Ryzin 1981). EPA has proposed the ED_{10} as a point of departure for cancer risk assessment (EPA 1996a).

It is recommended that, for chemicals for which there are sufficient dose-response data, a BMD corresponding to a 1% risk (BMD_{01}) be used instead of the NOAEL and that a BMD corresponding to a 10% risk (BMD_{10}) be used instead of the LOAEL. Like the NOAEL and LOAEL, the BMD_{01} and BMD_{10} are merely starting points for establishing safe exposures, but they have more precise definition and determination. Like the NOAEL and LOAEL, they are meant to correspond to very low risk. These BMDs should serve as starting points for setting acceptable human exposures to substances for *all* types of toxic effects, whether carcinogenic or noncarcinogenic. In the process of setting the exposure levels, BMDs must be modified by appropriate conversion factors and reduced by appropriate uncertainty factors, as will be discussed in subsequent sections. The resulting exposure guidance levels do not have specific risk connotations attached to them, but they are simply expected to reflect adequate safety.

When sufficient data are available, the unified BMD-based method for calculating acceptable human exposures is recommended for determining maximum contamination in water aboard spacecraft—spacecraft water exposure guidelines (SWEGs). The BMD approach is an evolving strategy that will assist in the calculation of SWEGs when adequate data are available. At the current stage of development of the BMD, the recommended method for establishing SWEGs is to determine the lower-confidence, model-based likelihood of a BMD_{01} level and apply appropriate uncertainty factors if necessary. This approach represents a recommended decision process to establish acceptable guidelines, but others, such as the lower confidence limit of a BMD_{10} or central estimates of the BMD, might be more useful for data that are available. Further evolution of BMD methodology should be monitored and appropriate alterations in the approach should be made as warranted. In the absence of sufficient data, or when special circumstances dictate, the recommended default procedure for determining SWEGs is essentially the NOAEL-based procedure currently in use for setting maximum contamination levels in air aboard spacecraft—spacecraft maximum allowable concentrations (SMACs) (NRC 1992; James and Gardner 1996).

BMD CALCULATION

Central Estimate Versus Confidence Limit

Estimating the BMD_p involves fitting a dose-response model to observed data and calculating the dose level that corresponds to an excess response, *p*, above background. Because the estimation of benchmark doses does not stray far from the observed data range, the choice of model might not be critical. As pointed out by Krewski and Van Ryzin (1981), fitted dose-response models for quantal toxicity data do not differ appreciably at responses above 1%. However, as much as possible, knowledge of the biologic mode of action should be used in modeling the dose-response data (Andersen et al. 2000; Wiltse and Dellarco 2000). Clearly, the validity of the observed dose-response data for risk assessment must be ascertained before BMD_p estimation begins.

It is desirable that methods used to fit dose-response models to observed data include provisions for calculating statistical confidence limits, because experimental variation is a source of uncertainty that must be considered. Instead of using a *formal* statistical lower confidence limit on a BMD_p as a starting point, one could calculate a central estimate of the BMD_p, and reduce it by an uncertainty factor to account for experimental variation. The result would be the same, but the expression of the lower confidence limit on the BMD_p via an uncertainty factor for experimental variation provides explicit information regarding the magnitude of this source of uncertainty. The uncertainty factor would be just one of several that would be used to reduce the central estimate to an acceptable exposure guidance level (T.B. Starr, TBS Associates, personal communication, 1997). Thus, the use of confidence limits instead of central estimates is intended to capture the experimental uncertainty rather than to provide "better" estimates. This topic is discussed further later in this chapter.

Estimating BMD_p for Various Toxic Effects

Traditional methods of dose-response modeling for binomially distributed random variables can be used to estimate carcinogenic effects and other quantal toxic responses (lethality, some mutagenic responses) for which subjects are assumed to respond independently from one another. Maximum likelihood estimation is a commonly accepted method of fitting a variety of mathematical dose-response models, including the multistage model, the probit model, or the Weibull model (Crump 1979; Zeise et al. 1987). The maximum likelihood method essentially identifies values of a model's parameters that have the highest likelihood of being correct, given the observed data used to fit the model. Generally, the only data available for modeling will be crude, lifetime incidences. If, however, data on time to occurrence of effects are available, the use of a model that can exploit this additional information is encouraged (e.g., Lensing and Kodell 1995).

Maximum likelihood estimation procedures have been worked out for fitting dose-response models for quantal effects that are assumed to be correlated between subjects. Chen and Kodell (1989), Ryan (1992), Allen et al. (1994a,b), and Krewski and Zhu (1995) all have proposed methodology for calculating BMD_p for toxicity data that are overdispersed with respect to simple binomial variation.

For continuous data, such as that arising in neurotoxicity studies, the definition of frank, adverse effects is not straightforward. Such data often are described well by normal (Gaussian) or lognormal distributions. Hence, modeling continuous responses on a probability scale to estimate the dose corresponding to a specified probability, p, of an adverse effect (BMD_p) is difficult. However, methods of risk assessment for such data have been developed (Gaylor and Slikker 1990; Kodell and West 1993; Crump 1995; Kavlock et al. 1995; Bosch et al. 1996), including provisions for calculating BMD. Hence, BMD_p for continuous, quantitative toxic responses can be calculated.

Appendix B provides examples of BMD estimation.

Exposure Conversion

Target Tissue Dose

Toxic substances sometimes require some form of metabolic activation to exert their adverse health effects, which might range from direct, short-term, target tissue toxicity to carcinogenesis. If metabolic activation can be characterized adequately by a pharmacokinetic model, then the dose delivered to the target should be used instead of the administered dose, for purposes of dose-response modeling to estimate BMD. Although the use of delivered dose rather than administered dose can be expected to lead to more accurate predictions of risk, pharmacokinetic modeling could actually lead to additional uncertainty, if physiologically based pharmacokinetic models with many parameters are used for tissue dosimetry (Farrar et al. 1989; Portier and Kaplan 1989). Hence, the question of whether to use target tissue dose or administered dose for dose-response modeling depends on the degree of confidence that can be placed in the pharmacokinetic model.

Differences in Duration

For toxic effects that are believed neither to accumulate nor to increase in adversity over time, a single exposure level for a toxicant can be used for SWEGs of different durations. However, for many toxic end points, an adjustment of the exposure will be required when extrapolating from one duration to another. Whenever possible, such extrapolation should use substance-specific, time response information, which can be in the form of an empirical mathematical relationship between exposure concentration and duration. For example, ten Berge et al. (1986) investigated the relationship between concentration and exposure time based on mortality data from 20 acute studies of locally and systemically acting inhalation toxicants. Using probit analysis, they found that the relationship $C^N \times T = K$ provided a good explanation of the relationship between concentration and duration. C is the concentration of the agent, T is the duration of exposure, and K is a constant. The value of N was generally greater than 1 and had an average value of approximately 1.8. (In fact, they found that the relationship $C^n \times T^m = k$ described the data well; the average value of n was approximately 3.5 and the average value of m was approximately 2.0. The expression $C^n \times T^m = k$ is actually equivalent to $C^N \times T = K$, where $N = n/m$ and $K = k^{1/m}$. Alternatively, the ten Berge rule could be expressed as $C \times T^M = K$, with $M = m/n$ and $K = k^{1/n}$.)

The simplest form of ten Berge's formula is $C \times T = K$, commonly known as Haber's rule, for inhalation toxicants. In the absence of chemical-specific information on the relationship between concentration and duration, Haber's rule often has been used as a default approach for making conversions for different (relatively short) durations of exposure. In its guidelines for the establishment of SMACs for airborne contaminants, the NRC (1992) urged caution in the use of this simple approach, and, for noncarcinogenic effects, the NRC subcommittee on SMACs has been reluctant to endorse its use for converting doses derived from longer term exposures to doses that would apply for shorter term exposures (see also James and Gardner 1996). However, like the NRC Subcommittee on Emergency Exposure Guidance Levels (NRC 1986), the subcommittee on SMACs does consider the use of $C \times T = K$ appropriate for extrapolating between two exposures that are relatively short term with respect to clearance or repair rate. Also, in the absence of definitive information, the subcommittee on SMACs has endorsed this approach for converting doses corresponding to shorter term exposures to doses corresponding to longer term exposures, although each substance must be considered individually with respect to the applicability of Haber's rule.

A simple comparison of ten Berge's rule to Haber's rule is given in Table 4-1, using $N = 2$, where the reference concentration is 50 parts per million (ppm) with a duration of 2 days (d). Conversions are made for 1-d and 4-d exposures. For $N > 1$, ten Berge's rule will give smaller concentrations than will Haber's rule in converting to shorter durations. It will give larger concentrations than will Haber's rule in converting to longer durations.

The National Advisory Committee for Acute Exposure Guideline Levels for Hazardous Substances (NAC/AEGL Committee) (EPA 1997) uses the relationship $C^N \times T = K$ proposed by ten Berge et al. (1986) to make conversions for different exposure durations in the setting of AEGLs. This relationship should be used whenever possible in making duration conversions when setting SWEGs for water contaminants

TABLE 4-1 Haber's and ten Berge's Rule Compared

Haber's Rule			ten Berge's Rule		
Concentration	Time	*K*	Concentration	Time	*K*
100 ppm	1 d	100	71 ppm	1 d	5000
50 ppm	2 d	100	50 ppm	2 d	5000
25 ppm	4 d	100	35 ppm	4 d	5000

Using 50 ppm as the reference concentration (exposure duration = 2 d; N = 2), Haber's rule and ten Berge's rule were used to make conversions for 1 d and 4 d exposures.

aboard spacecraft. As recommended by ten Berge et al. (1986), the value of N for specific toxicants should be derived empirically from experiments that provide data on various concentrations and various durations of exposure. This can be done by probit analysis. However, even when chemical-specific data are not available for estimating N, it might be possible to choose a default value of N other than $N = 1$ (Haber's rule), which would be expected to reflect the likely relationship between concentration and duration for a broad range of substances. The NAC/AEGL Committee (EPA 1997) often uses a default value of $N = 2$ when no exposure-versus-time data are available (e.g., arsine, 1,2-dichloroethane).

The method provided in the SMACs subcommittee's guidelines (NRC 1992) for converting lifetime daily exposure to carcinogens to exposures applicable to the shorter durations associated with spaceflight is based on a multistage model (Kodell et al. 1987) that is equivalent to using a $C \times T = K$ adjustment combined with an additional adjustment factor, f (NRC 1992). Murdoch et al. (1992) show that, for typical astronauts, f would not be likely to exceed a value of 2. In fact, for many plausible spaceflight scenarios, f will be about 1 (e.g., 3-stage model, first stage dose-related, and 30-year-old astronaut) for a wide range of exposure durations (e.g., 1-1000 d).

Species Conversions

Conversion of BMD_p values derived from data on an appropriately selected test species to comparable values for humans requires experienced scientific judgment. In the best situation, data on metabolism and disposition of the substance of interest in humans and the test species should be used to determine the appropriate conversion factor. For carcinogenic effects, interspecies conversions often are made on the basis of body weight or surface area differences between species (Allen et al. 1988; Travis and White 1988; EPA 1992). Such conversions are intended to correct for metabolic rate based on body size. However, their basis is related more to rates of basal metabolism than it is to xenobiotic metabolism (NRC 1992). Hence, quite often, in the absence of adequate pharmacokinetic or pharmacodynamic information to enable determination of an appropriate conversion factor, an assumption of concentration equivalence between species is made, and extrapolation is done on a straight concentration basis (e.g., parts per million in food, air, or water). Then an uncertainty factor generally is applied to account for unknown and unmeasured species differences.

Different Routes

In most cases, a BMD_p for SWEGs will be derived from oral exposure studies in animals. However, some might be based on nonoral routes. Exposures of humans during spaceflight to contaminated water can happen by a variety of routes: inhalation, water consumption, dermal absorption. Where possible, conversions must be made to account for differences between routes of exposure for astronauts and those used in the animal studies from which a BMD_p is derived. Assuming that the species-to-species conversion is made separately, all that would be required at this step would be a route-to-route conversion within species. At the very least, differences in rates of absorption for various routes should be considered where possible.

UNCERTAINTY FACTORS

Exposure Duration Uncertainty

When there is insufficient information available on a toxic substance to allow an informed adjustment to be made for differences in exposure duration, and when the rule $C^N \times T = K$ (ten Berge et al. 1986) cannot be applied even with a default value for N, it might be necessary to employ an uncertainty factor when extrapolating from one exposure duration to another. In many risk assessment exercises, such as in the derivation of RfDs, an exposure duration uncertainty factor is used when subchronic data must be used to set limits for chronic exposure (Barnes and Dourson 1988). In the past, the default value for this subchronic-to-chronic factor has been 10. In the setting of SWEGs, extrapolating from exposure durations for which data are available to those encountered in spaceflight might require the use of exposure duration uncertainty factors similar to the subchronic-to-chronic factor.

Interspecies Uncertainty

If sufficient pharmacokinetic and pharmacodynamic data are available, then a species-to-species conversion of the BMD_p should be made on as quantitative a basis as possible, using experienced scientific judgment. Unfortunately, such data are the exception rather than the rule. In most cases, there is an insufficient quantitative basis for making an informed extrapolation from animals to humans. Thus, it is considered prudent to reduce the BMD_p by an appropriate uncertainty factor to account for unknown species differences that might imply a greater sensitivity in humans than in experimental animals. Traditionally, a value of 10 has been used for this species factor, which originally arose in the setting of ADIs for chemicals in the food supply (NRC 1970). Factors greater or less than 10 should be used, depending on the nature of the toxicity. For example, central nervous system effects in most species might be similar to effects in humans, implying a species factor close to 1, whereas the uncertainty factor for other toxic effects might need to be as high as 15 (Calabrese and Baldwin 1995). The choice of a particular interspecies uncertainty factor needs to be justified in each case. A factor of 10 continues to be the default recommended by EPA (1996a), and it should be used as the default factor for SWEGs as well.

Experimental Variation

We recognize that the size of the confidence interval is highly dependent on the number of experimental subjects, often very small at lower doses. Rather than abandoning the BMD_{01} and using the central estimate, we suggest involving a modification of the uncertainty factor for small numbers. Pragmatically, this uncertainty factor would be used when even small numbers of subjects reflect little experimental variability. Examples include use of primates or other large test species where sample size may be limiting. The uncertainty factor would add 1% to 100% of $z_\alpha[(1 - p)/(np)]^{1/2}$ to the BMD_{01}, depending on certainty; z_α is the $100(1 - \alpha)^{th}$ percentile of a standard normal probability distribution, p is the excess response rate at the BMD_p, and n is the sample size on which p is based.

Experimental variation will be an important source of uncertainty in the derivation of SWEGs. Generally, sampling variation will be accounted for by using a statistical lower confidence limit on the BMD_p as an anchor point for setting a SWEG. However, it is not absolutely necessary that formal statistical confidence limits be used. Instead, an uncertainty factor for experimental variation could be included as one of several uncertainty factors used to reduce a central estimate of the BMD_p—we could call it the $CBMD_p$—to a SWEG. That is, as suggested by T.B. Starr (TBS Associates, personal communication, 1997), a formal statistical lower confidence limit on the BMD_p—say the $LBMD_p$—could be calculated and then used to back-calculate the appropriate factor, $f = CBMD_p/LBMD_p$, by which to reduce the central estimate of the BMD_p to account for experimental variation. Although that approach is exactly the same as calculating a formal statistical lower limit in the first place, it does have the advantage of conveying the size of the uncertainty factor, f, that is used to control for experimental variation. It should be noted that the size of f can be influenced both by the ability of the model to fit the observed dose-response relationship and by any constraints imposed as part of the fitting procedure.

In some situations, the available data might not permit the calculation of a statistical lower confidence limit on the BMD_p. This could happen, for example, when a maximum likelihood estimation procedure is used to calculate confidence limits, but, because of poor data, the procedure will not converge for the restricted, lower limit dose-response model. For such situations, R.L Kodell and D.W. Gaylor (National Center for Toxicological Research, Food and Drug Administration, unpublished material, 1998) have shown that an ad hoc uncertainty factor, f, for experimental variation can be calculated by

$$f = 1 + z_\alpha[(1 - p)/(np)]^{1/2},$$

where p is the excess response rate at the BMD_p, n is the sample size on which p is based, and z_α is the $100(1 - \alpha)^{th}$ percentile of a standard normal probability distribution ($z_\alpha = 1.645$ for 95% confidence).

The application of an uncertainty factor for sampling variation reflects the spirit of the "small-n" factor, $10/\sqrt{n}$, which is used by the NRC (1994) in the derivation of SMACs to reflect the uncertainty in NOAELs based on a limited number of human subjects. The small-n factor also could be used in the derivation of SWEGs if insufficient data are available to calculate a specific BMD_p.

BMD_{10} to BMD_{01}

If the BMD_{10} rather than the BMD_{01} is chosen as the anchor point for establishing a SWEG, which could happen—for example, if the estimated BMD_{01} is considered too unreliable or unstable—then it is logical to reduce the BMD_{10} by an additional uncertainty factor. An uncertainty factor of 10 is generally used to extrapolate from a LOAEL to a NOAEL. Because, in this document, the BMD_{01} is recommended in place of the NOAEL and the BMD_{10} is recommended in place of the LOAEL, it might be advisable to use the same factor of 10 to establish equivalence when extrapolating from a BMD_{10} to a BMD_{01}. However, some investigators indicate that a factor of 3-5 would be more appropriate for going from a LOAEL to a NOAEL (Abdel-Rahman and Kadry 1995), and the NRC subcommittee on SMACs endorsed the selective use by the National Aeronautics and Space Administration (NASA) of a factor of 2 for short-term irritation, based on available dose-response information (NRC 1994). In the examples of BMD calculation in Appendix B, all ratios of BMD_{10} to BMD_{01} lie between 2 and 10. It could be advantageous to use a BMD_{01} whenever possible rather than start with a BMD_{10} and have to reduce it as much as 10-fold. The recommendation here is that a factor of 3 or 10 be used to reduce a BMD_{10} to an appropriate BMD_{01} equivalent for setting SWEGs.

Environmental Effects

The special conditions of the space environment must be considered in defining SWEGs. Environmental factors that could alter the toxicity of water contaminants include microgravity, radiation, and stress (Kaplan 1979; Merrill et al. 1990). Astronauts can be physically, physiologically, and psychologically compromised in several ways: decreased muscle mass, decreased bone mass, decreased red-blood-cell mass, depressed immune systems, altered nutritional requirements, behavioral changes, shift of body fluids, altered blood flow, altered hormonal status, altered enzyme concentrations, increased sensitization to cardiac arrhythmia, and altered drug metabolism (NRC 1992). Hence, astronauts in space will be in an altered homeostatic state and might experience increased sensitivity to the toxic effects of contaminated water.

It is important to reduce chemical exposure relative to what would be acceptable on Earth for toxic effects that are influenced by the physiologic changes induced by spaceflight. However, there is generally little definitive information to permit a precise, quantitative conversion that would reflect altered toxicity resulting from spaceflight environmental factors. Hence, the use of an uncertainty factor generally is dictated when available information on a substance indicates that it affects one or more aspects of an astronaut's condition that might be compromised in space. For example, the SMACs subcommittee has agreed with NASA's practice of applying an uncertainty factor of 3 or 5 to modify allowable exposure concentrations for chemical agents that affect the immune system or that have been demonstrated to sensitize animals to cardiac arrhythmia (NRC 1992, 1994, 1996a,b; James and Gardner 1996).

When data on the effects of microgravity on bodily functions are available, information derived from them might preclude the need for an uncertainty factor. For example, in a study of pulmonary function of astronauts who participated in flights lasting 9-14 d aboard NASA's *Spacelab*, West et al. (1997) conclude that, although there were adaptive changes in pulmonary function in microgravity, none of the observed changes would limit spaceflight. Based on this observation, an uncertainty factor for microgravity for pulmonary toxicants under these exposure situations (i.e., as in the West study) might not be required for flights of short duration. For prolonged exposure in space, the practice of using an uncertainty factor might be warranted, but it can be revised, as human data (or suitably predictive animal data) become available.

Other Factors

The conversion factors and uncertainty factors represent the generic modifying factors that must be applied routinely in deriving SWEGs from BMD_ps. There could be additional modifying factors that should be applied in specific cases, depending on the substance in question and the nature of the space mission. For example, one important uncertainty factor commonly used in risk assessments that concern general public health is a factor to account for variability among humans in sensitivity to specific substances. Because of the relatively homogeneous, robust health status of astronauts on most space missions, it is not necessary routinely to apply an uncertainty factor for intraspecies variability. To date, biologic diversity (age, sex, toxicogenetic differences) has not presented significant concern. However, there might be special missions or specific substances for which use of an intraspecies factor would be warranted. This issue would be considered case by case, and it is beyond the scope of this document. When the Human Genome Project and the Environmental Genome Project are completed, data will begin to accumulate on genetic polymorphisms. NASA should monitor the progress in the identification of polymorphisms that make certain individuals more susceptible to certain chemicals.

Another modifying factor sometimes applied in general risk assessment practice is one that reflects uncertainty about the quality of data. That is, if the data are considered inadequate or incomplete, an allowable exposure level would be reduced by some factor. However, it has been the practice of the NRC subcommittee on SMACs to recommend that SMACs not be established for substances for which the data are inadequate, rather than set an unreasonably low SMAC that would give the appearance of being data-based when it was not (NRC 1994, 1996a,b). This practice is recommended for setting SWEGs.

Although the intent of the procedure recommended for setting SWEGs is to provide a unified approach that applies to all types of toxic effects, there is one possible point of departure between carcinogenic and noncarcinogenic effects. That is, because of the severity and irreversibility of cancer, some risk assessors recommend that exposure limits based on carcinogenic effects be reduced by an additional factor to take this into account (Renwick 1995; Gaylor et al. 1999). The subcommittee recommends that this issue be considered on a case-by-case basis, and that any use of an additional uncertainty factor be scientifically justified.

SETTING SWEGs

Overall Uncertainty Factor

Each uncertainty factor discussed earlier accounts for one source of uncertainty for which either the direction of the difference or the magnitude of the difference between an estimated value and the true value is unknown. Each factor is presumed to account for extreme differences that might exist between estimated and true values. Because not all true differences would be expected to be at their extremes simultaneously, reducing a BMD_p by a *product* of uncertainty factors could lead to undue conservatism—in the sense that the resulting SWEG might be lower than necessary to provide the desired protection. Recognizing the compounding of conservatism that occurs in dividing experimental doses by multiple uncertainty factors, as demonstrated by Bogen (1994) and Slob (1994), Gaylor and Chen (1996) suggest using a reduced, combined uncertainty factor to set acceptable limits. Assuming that individual uncertainty factors are lognormally distributed (Dourson et al. 1996), a combined uncertainty factor, F, can be calculated as follows (Kodell and Gaylor 1999; Gaylor and Kodell 2000):

$$F = \exp\{\Sigma_i \mathrm{avg}[\ln(f_i)] + z_\alpha(\Sigma_i s^2_{\ln(fi)})^{1/2}\},$$

where avg[ln(f_i)] is an estimate of the mean $\log_e$-uncertainty factor for the ith of m sources of uncertainty, $s_{\ln(fi)}$ is an estimate of the standard deviation of the distribution of ln(f_i), and z_α is the $100(1 - \alpha)^{th}$ percentile of the standard normal distribution.

Implicit in the use of individual uncertainty factors is the assumption that true conversion factors for the various types of extrapolation are random variables, and that the individual uncertainty factors capture a high percentage of the range of variation for each extrapolation. A factor of 10 is the default value for most factors, but some investigators argue for larger or smaller factors for specific sources of uncertainty. For example, Swartout (1996) observed that a factor as large as 17 might be necessary to cover the uncertainty in estimating chronic effects using subchronic data. Abdel-Rahman and Kadry (1995) argue that a factor as small as 3 could be sufficient to capture LOAEL-to-NOAEL uncertainty (and perhaps also BMD_{10}-to-BMD_{01} uncertainty).

The rationale for the combined uncertainty factor *F* is that, if estimates of the mean and standard deviation of the individual distributions of uncertainty are available, then statistical techniques for estimating upper tolerance limits of distributions of sums of independent random variables can be used to calculate a reduced overall uncertainty factor (that is less than the product of individual factors) that will still capture a high percentage of the overall range of uncertainty. Specifically, the formula for *F* is a point estimate of the $100(1 - \alpha)^{th}$ percentile (e.g., 95th percentile, for $\alpha = 0.05$) of the combined range of uncertainty. The use of this combined factor would be expected to provide $100(1 - \alpha)\%$ assurance of protection for the combined sources of uncertainty. Recent studies by Baird et al. (1996) and Swartout et al. (1998) have used Monte Carlo simulation techniques to estimate upper percentiles of the distribution of combined uncertainty factors (simulation-based values of *F*) for consideration in the setting of RfDs. The combined uncertainty factor *F* is recommended for consideration and use in the process of setting SWEGs.

Table 4-2 contains information derived from the literature on averages and standard deviations of $\log_e$-uncertainty factors for sources of uncertainty that are commonly encountered in risk assessment extrapolations.

TABLE 4-2 Estimated Averages and Standard Deviations of Log_e-Uncertainty Factors ($\ln(f_i)$) for Various Sources of Uncertainty

Source of Uncertainty	Average	($\ln(f_i)$)	Reference
Human-to-human	0	1.64	Dourson and Stara 1983
Animal-to-human	0	1.66	Calabrese and Baldwin (1995)
Subchronic-to-chronic	0.69	1.30	Swartout 1996
LOAEL-to-NOAEL	1.25	0.60	Abdel-Rahman and Kadry 1995

Based on NASA's experience in setting SMACs for air contaminants, it appears that practical application of the combined uncertainty factor in setting SWEGs for water contaminants will generally involve only a pair of individual uncertainty factors, one for extrapolating between species and one for extrapolating from a BMD_{10} to a BMD_{01} (or LOAEL to NOAEL). However, an exposure duration uncertainty factor, such as the subchronic-to-chronic factor, might be needed for spaceflights of 1000 d. In general, a factor to account for human variability will not be necessary in the calculation of SWEGs, because variability among individual astronauts would be expected to be small relative to the general population. Based on the NRC formula applied to data from Table 4-2, the upper 95% tolerance limit for animal-to-human variability alone would be

$$F = \exp[0 + (1.645)(1.66)] \doteq 15.$$

This is near the customary default uncertainty factor of 10. The upper 95% tolerance limit for LOAEL-to-NOAEL (BMD_{10}-to-BMD_{01}) uncertainty alone would be

$$F = \exp[1.25 + (1.645)(0.60)] \doteq 10.$$

This is the common default value for this source of uncertainty. Combining the two sources of uncertainty via the above formula gives a combined uncertainty factor (an estimated upper 95% tolerance limit) of

$$F = \exp\{0 + 1.25 + 1.645[(1.66)^2 + (0.60)^2]^{1/2}\} \doteq 64.$$

Hence, instead of the product of factors, i.e., 15 × 10 = 150 (or, commonly, 10 × 10 = 100), a reduced factor of 64 can be used with 95% assurance of capturing these two sources of uncertainty. By comparison, the standard product of defaults (10 × 10 = 100) gives about 97% assurance.

The formula for *F* is intended to apply only to uncertainty factors, and only to those for which estimates of the mean and standard deviation of the distribution are available. It does not apply to exposure conversion factors, factors that account for severity, factors that account for additive or synergistic effects of space flight, factors that account for data inadequacy, the uncertainty factor for experimental variation, or uncertainty factors for which distributional information is unavailable. All of those factors must continue to be applied separately.

Mixtures of Chemicals

SWEGs for single toxic constituents are set individually, without regard to their occurrence in mixtures with other chemicals in spacecraft water. However, if a substance is present in water, its presence will always be as one component of a complex mixture. Therefore, individual SWEGs must be integrated into group limits to reflect overall water quality conditions judged to be safe for humans in space flight. Those substances that have similar modes of action, or that induce effects in a particular target organ, and might be assumed concentration-additive or perhaps synergistic, could be grouped together, and their respective concentrations, C_i, could be determined as follows (NRC 1987a):

$$\frac{C_1}{SWEG_1} + \frac{C_2}{SWEG_2} + \frac{C_k}{SWEG_k} \leq 1.$$

C is the measured concentration of a particular chemical in spacecraft water, which is divided by the corresponding SWEG for that chemical.

The group-limit concept has been endorsed by the American Conference of Governmental Industrial Hygienists (ACGIH) for chemical concentrations in air (ACGIH 1991) and by EPA for chemical mixtures that occur in any exposure medium (EPA 1986). For each group with a particular mode of action, a separate group limit calculation should be made for restricting the concentrations of these species in spacecraft water. If it is known or suspected that the action of a mixture of chemicals is greater than additive, then the group limit concept will not guarantee protection. In that case, a further restriction of concentrations is warranted.

Multiple Toxic End Points

In general, toxic substances can affect more than one organ system and have more than one effect within an organ system. In the setting of SWEGs, all observed toxic effects are considered, including mortality, morbidity (functional impairment), reproductive toxicity, genotoxicity, carcinogenicity, neurotoxicity, immunotoxicity, hepatotoxicity, and respiratory toxicity. For effects that are considered relevant to humans, generally the most sensitive effect determines the SWEG. That is, for a given chemical, potential SWEGs are calculated for a specific exposure duration in space (1, 10, 100, or 1000 d) using data on a variety of toxic end points. Based on the critical significance of the health effect identified, the lowest of those potential SWEGs is then chosen as the SWEG for that substance for that duration of exposure. However, any potential SWEG that is within a factor of 3 of the lowest potential SWEG is also considered a determining factor of that SWEG.

Comparisons with Established Values

All documents used to establish previous industrial or public-health exposure guidance levels for water contaminants should be reviewed before SWEG values are set for NASA. In particular, previous NRC documents on acceptable exposures in drinking water (e.g., NRC 1987b) and water contaminant limits established by EPA, both the maximum contaminant limits and the health advisories (EPA 1996b), provide important reference points for comparison. Such comparisons are not simply to mimic guidance levels set by other entities, but to determine whether the SWEGs that are set in response to NASA's special needs are reasonable in light of previously set concentrations. Finally, the NRC documents on SMACs must be reviewed to ensure compatibility of standards for water with those for air (NRC 1994; 1996a,b; 2000). Any significant differences between exposure levels should be discussed and justified, which would include an evaluation of the approaches and data used to derive the guidance levels.

Alternative Approaches

This section describes several approaches that are under development for setting acceptable exposure guidelines for toxic substances. These approaches are valid to consider when setting SWEGs, although they generally require more data than are readily available. Progress in the development of these refined approaches should be monitored, so that they can be included as appropriate in the process of setting SWEGs.

Integrated PBPK and BBDR Models

Just as the multistage model of Armitage and Doll (1960) replaced the earliest susceptibility models, probit and logit, as the primary basis for carcinogenic risk assessment, in recent years more refined biologically based dose-response (BBDR) models have been proposed as replacements for the multistage model. The most popular is the two-stage clonal expansion model of Moolgavkar and colleagues (e.g., Moolgavkar and Luebeck 1990), and extensions thereof (Portier and Kopp-Schneider 1991; Zheng et al. 1995). The BBDR model characterizes the important role of cellular proliferation in cancer. There has been a strong belief that, with sufficient biologic data on the components of the cancer process, complex BBDR models will provide the means for estimating risks below the dose-response range based on biologic knowledge rather than on assumptions.

Concomitant with the refinement of BBDR models has been the development of ever more sophisticated physiologically based pharmacokinetic (PBPK) models, which have been used to obtain better estimates of target tissue doses for risk assessment (Andersen et al. 1993; Kohn et al. 1993). In some cases, this has led to significant modification of risk assessments originally based on the linearized multistage model (e.g., Starr 1990). However, it is only recently that PBPK and BBDR models have been fully integrated into the risk assessment process. Although the results are still the subject of scientific debate, the recent reassessment of TCDD (2,3,7,8-tetrachlorodibenzo-*p*-dioxin) by EPA (1998) used a fully integrated PBPK-BBDR model to estimate risk of liver cancer in rats. Whether this refined approach will provide more reliable estimates of cancer risk below the experimental dose range is open to question. Nevertheless, the results should be carefully scrutinized to determine whether that is the appropriate direction for risk assessment to take. In the rare case that such data are available for this refined modeling, the exercise certainly should be carried out, and the results should be compared against the general procedure outlined above for setting SWEGs, before SWEG are set.

In addition to more refined biologic models for carcinogenesis, there have been isolated attempts to develop BBDR models for noncancer effects, specifically for developmental toxicity (Freni and Zapisek 1991; Shuey et al. 1994, 1995; Leroux et al. 1996) and neurotoxicity (Slikker et al. 1998).

Ordinal Regression

A technique called ordinal regression has been proposed as a way to combine data on various toxic effects into a single analysis. The method was first proposed by Hertzberg and Miller (1985) and was later refined by Guth et al. (1991). With ordinal regression, health effects are first assigned to severity categories based on the reported information and consideration of biologic and statistical significance. The aggregate group of subjects at any particular dose and duration of exposure is classified as giving evidence of a specific severity of response. Models such as the logistic regression model are applied with the severity code as the dependent variable and the exposure concentration, duration of exposure, and species as the independent variables. The method allows incorporation of quantal and quantitative data, and it enables the simultaneous analysis of data from many studies. One tradeoff is the loss of target-organ toxicity.

The output from ordinal regression is especially useful in that, for any level of severity, it can provide a concentration-by-duration profile (central estimates and confidence limit estimates) for any amount of risk. That capability is particularly useful for making duration conversions, because approaches such as the concentration-by-time conversion are not required. Furthermore, if sufficient human data are available to include in the regression analysis, then interspecies uncertainty is reduced. The ordinal regression method continues to be refined and to be applied to specific toxicants (Simpson et al. 1996), but the complexities of the model fitting appear to make it infeasible for routine use. Nevertheless, whenever possible, the method ought to be applied, and the results should be compared against the general procedure outlined above for obtaining SWEGs before SWEG values are set. Most if not all applications of ordinal regression have been restricted to acute toxic effects, specifically excluding carcinogenic effects. Whether all types of toxic effects, including cancer, can be modeled simultaneously using ordinal regression is still undetermined.

Change-Point Dose-Response Models

In a practical sense, replacing a NOAEL with a BMD_p is scientifically justified, in light of the studies reported above that have documented nonzero risk (risk = $p > 0$, for $0.01 \leq p \leq 0.10$) associated with NOAELs. In a theoretical sense, however, replacing a NOAEL with a BMD_p is inconsistent, because a NOAEL is intended to represent a dose with true zero risk—a threshold. In theory, then, if it were possible to estimate reliably a true threshold dose by way of a mathematical model, it would make sense to replace the NOAEL with this estimated threshold dose, instead of the BMD_p. There is research into the use of so-called change-point dose-response models for risk assessment, where the change-point is a dose value that determines where the model changes from a constant response model to a dose response model. Hence, the change-point is a threshold dose, which is a parameter estimated as part of the modeling exercise. It must be emphasized that any estimate of a threshold from current BMD dose-response models is entirely empirical and has no biological basis. If change-point models are shown to be practical, then the guidelines for basing SWEGs on BMD_ps should be revisited to evaluate the feasibility of using change-points instead of BMD_ps for presumed threshold effects. However, the use of estimated change-points for threshold effects and BMD_ps for nonthreshold effects would destroy the unity of the proposed approach for all types of toxic effects, including threshold and nonthreshold effects.

SUMMARY

Using the process of risk assessment to establish SWEGs involves several important steps. Although the intent here has been to provide guidance for implementing this step-by-step process, it must be emphasized that scientific judgment is critical at every step, and it should be the overriding factor throughout the process. Scientific judgment is based on the aggregate of biologic information. Because the process involves a series of extrapolations, each with its own degree of uncertainty, attempting to identify exposures to which specific, low amounts of risk can be attached is not recommended. Instead, emphasis is placed on establishing concentrations that are judged to be reasonably safe for human exposure, based on the best scientific information and judgment available. This approach to establishing SWEGs is in line with current thinking on risk assessment, which is moving away from emphasizing numerical estimates of risk for extremely low exposures and is moving toward simply identifying exposures for which the risk of adverse human health effects is judged to be negligible, regardless of whether such effects are carcinogenic or not.

REFERENCES

Abdel-Rahman, M.S., and A.M. Kadry. 1995. Studies on the use of uncertainty factors in deriving RfDs. Hum. Ecol. Risk Assess. 1(5):614-624.

ACGIH (American Conference of Governmental Industrial Hygienists). 1991. Documentation of the Threshold Limit Values and Biological Exposure Indices, 6th Ed. Cincinnati, OH: American Conference of Governmental Industrial Hygienists.

Allen, B.C., K.S. Crump, and A.M. Shipp. 1988. Correlation between carcinogenic potency of chemicals in animals and humans. Risk Anal. 8(4):531-544.

Allen, B.C., R.J. Kavlock, C.A. Kimmel, and E.M. Faustman. 1994a. Dose-response assessment for developmental toxicity. II. Comparison of generic benchmark dose estimates with no observed adverse effect levels. Fundam. Appl. Toxicol. 23(4):487-495.

Allen, B.C., R.J. Kavlock, C.A. Kimmel, and E.M. Faustman. 1994b. Dose-response assessment for developmental toxicity. III. Statistical models. Fundam. Appl. Toxicol. 23(4):496-509.

Andersen, M.E., J.J. Mills, M.L. Gargas, L. Kedderis, L.S. Birnbaum, D. Neubert, and W.F. Greenlee. 1993. Modeling receptor-mediated processes with dioxin: Implications for pharmacokinetics and risk assessment. Risk Anal. 13(1):25-36.

Andersen, M.E., M.E. Meek, G.A. Boorman, D.J. Brusick, S.M. Cohen, Y.P. Dragan, C.B. Frederick, J.I. Goodman, G.C. Hard, E.J. O'Flaherty, and D.E. Robinson. 2000. Lessons learned in applying the U.S. EPA proposed cancer guidelines to specific compounds. Toxicol. Sci. 53(2):159-172.

Armitage, P., and R. Doll. 1960. Stochastic models for carcinogenesis. Pp. 19-38 in Proceedings of the Fourth Berkeley Symposium on Mathematical Statistics and Probability, Vol. 4, J. Neyman, ed. Berkeley, CA: University of California Press.

Baird, S.J.S., J.T. Cohen, J.D. Graham, A.I. Shlyakhter, and J.S. Evans. 1996. Noncancer risk assessment: A probabilistic alternative to current practice. Hum. Ecol. Risk Assess. 2(1):79-102.

Barnes, D.G., and M. Dourson. 1988. Reference dose (RfD): Description and use in health risk assessments. Regul. Toxicol. Pharmacol. 8(4):471-486.

Bogen, K.T. 1994. A note on compounded conservatism. Risk Anal. 14(4):379-381.

Bosch, R.J., D. Wypij, and L.M. Ryan. 1996. A semiparametric approach to risk assessment for quantitative outcomes. Risk Anal. 16(5):657-665.

Calabrese, E.J., and L.A. Baldwin. 1995. A toxicological basis to derive generic interspecies uncertainty factors for application in human and ecological risk assessment. Hum. Ecol. Risk Assess. 1(5):555-564.

Chen, J.J., and R.L. Kodell. 1989. Quantitative risk assessment for teratological effects. J. Am. Stat. Assoc. 84(408):966-971.

Crump, K.S. 1979. Dose response problems in carcinogenesis. Biometrics 35(1):157-167.

Crump, K.S. 1984. A new method for determining allowable daily intakes. Fundam. Appl. Toxicol. 4(5):854-871.

Crump, K.S. 1995. Calculation of benchmark doses from continuous data. Risk Anal. 15(1):79-89.

Crump, K.S., D.G. Hoel, C.H. Langley, and R. Peto. 1976. Fundamental carcinogenic processes and their implications for low dose risk assessment. Cancer Res. 36(9 Pt. 1):2973-2979.

Crump, K.S., H.J. Clewell, and M.E. Andersen. 1996. Cancer and non-cancer risk assessment should be harmonized. BELLE Newslett. 5(2):2-4.

Dourson, M.L. 1994. Methods for establishing oral reference doses (RfDs). Pp. 51-61 in Risk Assessment of Essential Elements, W. Mertz, C.O. Abernathy, and S.S. Olin, eds. Washington, DC: ILSI Press.

Dourson, M.L., and J.F. Stara. 1983. Regulatory history and experimental support of uncertainty (safety) factors. Regul. Toxicol. Pharmacol. 3(3):224-238.

Dourson, M.L., S.P. Felter, and D. Robinson. 1996. Evolution of science-based uncertainty factors in noncancer risk assessment. Regul. Toxicol. Pharmacol. 24(2 Pt. 1):108-120.

EPA (U.S. Environmental Protection Agency). 1986. Guidelines for the Health Risk Assessment of Chemical Mixtures. Fed. Regist. 51(185):34014-34025.

EPA (U.S. Environmental Protection Agency). 1992. Draft report: A Cross-Species Scaling Factor for Carcinogen Risk Assessment Based on Equivalence of $mg/kg^{3/4}/day$. Fed. Regist. 57(109):24152-24173.

EPA (U.S. Environmental Protection Agency). 1996a. Proposed Guidelines for Carcinogen Risk Assessment. Fed. Regist. 61(79):17995-18011.

EPA (U.S. Environmental Protection Agency). 1996b. Drinking Water Regulations and Health Advisories. EPA-822-B-96-002. Office of Water, U.S. Environmental Protection Agency, Washington, DC.

EPA (U.S. Environmental Protection Agency). 1997. National Advisory Committee for Acute Exposure Guideline Levels for Hazardous Substances. Fed. Regist. 62(210):58839-58851.

EPA (U.S. Environmental Protection Agency). 1998. Dose-response modeling for 2,3,7,8-TCDD. Chapter 8 in Health Assessment Document for 2,3,7,8-Tetrachlorodibenzo-*p*-dioxin (TCDD) and Related Compounds. EPA/600/P-92/001C8. [Online]. Available: http://www.epa.gov/ordntrnt/ORD/WebPubs/dioxin/index.html

Farmer, J.H., R.L. Kodell, and D.W. Gaylor. 1982. Estimation and extrapolation of tumor probabilities from a mouse bioassay with survival/sacrifice components. Risk Anal. 2(1):27-34.

Farrar, D., B. Allen, K. Crump, and A. Shipp. 1989. Evaluation of uncertainty in input parameters to pharmacokinetic models and the resulting uncertainty in output. Toxicol. Lett. 49(2-3):371-385.

Freni, S.C., and W.F. Zapisek. 1991. Biologic basis for a risk assessment model for cleft palate. Cleft Palate Craniofac. J. 28(4):338-346.

Gaylor, D.W. 1983. The use of safety factors for controlling risk. J. Toxicol. Environ. Health 11(3):329-336.

Gaylor, D.W. 1992. Incidence of developmental defects at the no observed adverse effect level (NOAEL). Regul. Toxicol. Pharmacol. 15(2 Pt. 1):151-160.

Gaylor, D.W., and J.J. Chen. 1996. A simple upper limit for the sum of the risks of the components in a mixture. Risk Anal. 16(3):395-398.

Gaylor, D.W., and R.L. Kodell. 2000. Percentiles of the product of uncertainty factors for establishing probabilistic reference doses. Risk Anal. 20(2):245-250.

Gaylor, D.W., and W. Slikker, Jr. 1990. Risk assessment for neurotoxic effects. Neurotoxicology 11(2):211-218.

Gaylor, D.W., R.L. Kodell, J.J. Chen, J.A. Springer, R.J. Lorentzen, and R.J. Scheuplein. 1994. Point estimates of cancer risk at low doses. Risk Anal. 4(5):843-850.

Gaylor, D.W., R.L. Kodell, J.J. Chen, and D. Krewski. 1999. A unified approach to risk assessment for cancer and noncancer endpoints based on benchmark doses and uncertainty/safety factors. Regul. Toxicol. Pharmacol. 29(2 Pt. 1):151-157.

Guth, D.J., A.M. Jarabek, L. Wymer, and R.D. Hertzberg. 1991. Evaluation of Risk Assessment Methods for Short-term Inhalation Exposure. Abstract 91-173.2 in Proceedings of the 84th Annual Meeting of the Air and Waste Management Association. Pittsburgh, PA: Air & Waste Management Association.

Hertzberg, R.C., and M. Miller. 1985. A statistical model for species extrapolation using categorical response data. Toxicol. Ind. Health 1(4):43-57.

Hill, R.N., T.M. Crisp, P.M. Hurley, S.L. Rosenthal, and D.V. Singh. 1998. Risk assessment of thyroid follicular cell tumors. Environ. Health Perspect. 106(8):447-457.

James, J.T., and D.E. Gardner. 1996. Exposure limits for airborne contaminants in spacecraft atmospheres. Appl. Occup. Environ. Hyg. 11(12):1424-1425.

Kaplan, H.L. 1979. Contaminants. In The Physiological Basis for Spacecraft Environmental Limits. J.M. Waligora, coordinator. NASA-RP-1045. NASA Center for AeroSpace Information, Hanover, MD.

Kavlock, R.J., B.C. Allen, E.M. Faustman, and C.A. Kimmel. 1995. Dose-response assessments for developmental toxicity. IV. Benchmark doses for fetal weight changes. Fundam. Appl. Toxicol. 26(2):211-222.

Kimmel, C.A., and E.W. Gaylor. 1988. Issues in qualitative and quantitative risk analysis for developmental toxicology. Risk Anal. 8(1):15-20.

Klaassen, C.D., and D.L. Eaton. 1991. Principles of toxicology. Pp. 12-49 in Casarett and Doull's Toxicology: The Basic Science of Poisons, 4th Ed., M.O. Amdur, J. Doull, and C.D. Klaassen, eds. New York: Macmillan.

Kodell, R.L., and D.W. Gaylor. 1999. Combining uncertainty factors in deriving human exposure levels of noncarcinogenic toxicants. Ann. N.Y. Acad. Sci. 895:188-195.

Kodell, R.L., and C.N. Park. 1995. Linear extrapolation in cancer risk assessment. Pp. 87-104 in Low-Dose Extrapolation of Cancer Risks: Issues and Perspectives, W. Farland, S. Olin, C. Park, L. Rhomberg, R. Scheuplein, T. Starr, and J. Wilson, eds. Washington, DC: ILSI Press.

Kodell, R.L., and R.W. West. 1993. Upper confidence limits on excess risk for quantitative responses. Risk Anal. 13(2):177-182.

Kodell, R.L., D.W. Gaylor, and J.J. Chen. 1987. Using average lifetime dose rate for intermittent exposures to carcinogens. Risk Anal. 7:339-345.

Kohn, M.C., G.W. Lucier, G.C. Clark, C. Sewall, A.M. Tritscher, and C.J. Portier. 1993. A mechanistic model of effects of dioxin on gene expression in the rat liver. Toxicol. Appl. Pharmacol. 120(1):138-154.

Krewski, D., and J. Van Ryzin. 1981. Dose response models for quantal response toxicity data. Pp. 201-231 in Statistics and Related Topics: Proceedings of the International Symposium on Statistics and Related Topcis, M. Csorgo, D.A. Dawson, J.N.K. Rao, and A.K.Md.E. Saleh, eds. Amsterdam: North-Holland.

Krewski, D., and Y. Zhu. 1995. A simple data transformation for estimating benchmark doses in developmental toxicity experiments. Risk Anal. 15(1):29-39.

Lehman, A.J., and O.G. Fitzhugh. 1954. 100-fold margin of safety. Assoc. Food Drug Off., Q. Bull. 18:33-35.

Leisenring, W., and L. Ryan. 1992. Statistical properties of the NOAEL. Regul. Toxicol. Pharmacol. 15(2 Pt. 1):161-171.

Lensing, S.Y., and R.L. Kodell. 1995. Fitting the two-stage clonal expansion model based on exact hazard to the ED_{01} data using SAS NLIN. Risk Anal. 15(2):233-245.

Leroux, B.G., W.M. Leisenring, S.H. Moolgavkar, and E.M. Faustman. 1996. A biologically-based dose-response model for developmental toxicology. Risk Anal. 16(4):449-458.

Mantel, N., and W.R. Bryan. 1961. "Safety" testing of carcinogenic agents. J. Natl. Cancer Inst. 27(2):455-470.

Merrill, A.H., Jr., M. Hoel, E. Wang, R.E. Mullins, J.L. Hargrove, D.P. Jones, and I.A. Popova. 1990. Altered carbohydrate, lipid, and xenobiotic metabolism by liver from rats flown on Cosmos 1887. FASEB J. 4:95-100; erratum FASEB J. (1990) 4(8):2539.

Moolgavkar, S.H., and G. Luebeck. 1990. Two-event model for carcinogenesis: Biological, mathematical, and statistical considerations. Risk Anal. 10(2): 323-341.

Munro, I.C., and D.R. Krewski. 1981. Risk assessment and regulatory decision making. Food Cosmet. Toxicol. 19(5):549-560.

Murdoch, D.J., D. Krewski, and J. Wargo. 1992. Cancer risk assessment with intermittent exposure. Risk Anal. 12(4):569-577.

NRC (National Research Council). 1970. Evaluating the Safety of Food Chemicals: Report of the Food Protection Committee. Washington, DC: National Academy of Sciences.

NRC (National Research Council). 1986. Criteria and Methods for Preparing Emergency Exposure Guidance Level (EEGL), Short-Term Public Emergency Guidance Level (SPEGL), and Continuous Exposure Guidance Level (CEGL) Documents. Washington, DC: National Academy Press.

NRC (National Research Council). 1987a. Drinking Water and Health, Vol. 7—Disinfectants and Disinfectant By-Products. Washington, DC: National Academy Press.

NRC (National Research Council). 1987b. Drinking Water and Health, Vol. 8—Pharmacokinetics in Risk Assessment. Washington, DC: National Academy Press.

NRC (National Research Council). 1992. Guidelines for Developing Spacecraft Maximum Allowable Concentrations for Space Station Contaminants. Washington, DC: National Academy Press.

NRC (National Research Council). 1994. Spacecraft Maximum Allowable Concentrations for Selected Airborne Contaminants, Vol. 1. Washington, DC: National Academy Press.

NRC (National Research Council). 1996a. Spacecraft Maximum Allowable Concentrations for Selected Airborne Contaminants, Vol. 2. Washington, DC: National Academy Press.

NRC (National Research Council). 1996b. Spacecraft Maximum Allowable Concentrations for Selected Airborne Contaminants, Vol. 3. Washington, DC: National Academy Press.

NRC (National Research Council). 2000. Spacecraft Maximum Allowable Concentrations for Selected Airborne Contaminants, Vol. 4. Washington, DC: National Academy Press.

OSTP (Office of Science and Technology Policy). 1985. Chemical Carcinogens: A Review of the Science and Its Associated Principles. Fed. Regist. 50(50):10371-10442.

Page, N.P., D.V. Singh, W. Farland, J.I. Goodman, R.B. Conolly, M.E. Andersen, H.J. Clewell, C.B. Frederick, H. Yamasaki, and G. Lucier. 1997. Implementation of EPA revised cancer risk assessment guidelines: Incorporation of mechanistic and pharmacokinetic data. Fundam. Appl. Toxicol. 37(1):16-36.

Portier, C.J., and N.L. Kaplan. 1989. Variability of safe dose estimates when using complicated models of the carcinogenic process. A case study: Methylene chloride. Fundam. Appl. Toxicol. 13(3):533-544.

Portier, C.J., and A. Kopp-Schneider. 1991. A multistage model of carcinogenesis incorporating DNA damage and repair. Risk Anal. 11(3): 535-543.

Purchase, I.F., and T.R. Auton. 1995. Thresholds in chemical carcinogenesis. Regul. Toxicol. Pharmacol. 22(3):199-205.

Renwick, A.G. 1995. The use of an additional safety or uncertainty factor for nature of toxicity in the estimation of acceptable daily intake and tolerable daily intake values. Regul. Toxicol. Pharmacol. 22(3):250-261.

Ryan, L. 1992. Quantitative risk assessment for developmental toxicity. Biometrics 48(1):163-174.

Simpson, D.G., R.J. Carroll, H. Zhou, and D.J. Guth. 1996. Interval censoring and marginal analysis in ordinal regression. J. Agri. Biol. Environ. Stat. 1:354-376.

Shuey, D.L., C. Lau, T.R. Logsdon, R.M. Zucker, K.H. Elstein, M.G. Narotsky, R.W. Setzer, R.J. Kavlock, and J.M. Rogers. 1994. Biologically based dose-response modeling in developmental toxicology: Biochemical and cellular sequelae of 5-fluorouracil exposure in the developing rat. Toxicol. Appl. Pharmacol. 126(1):129-144.

Shuey, D.L., R.W. Setzer, C. Lau, R.M. Zucker, K.H. Elstein, M.G. Narotsky, R.J. Kavlock, and J.M. Rogers. 1995. Biological modeling of 5-fluorouracil developmental toxicity. Toxicology 102(1-2):207-213.

Slikker, W., Jr., A.C. Scallet, and D.W. Gaylor. 1998. Biologically-based dose-response model for neurotoxicity risk assessment. Toxicol. Lett. 102-103:429-433.

Slob, W. 1994. Uncertainty analysis in multiplicative models. Risk Anal. 14(4):571-576.

Starr, T.B. 1990. Quantitative cancer risk estimation for formaldehyde. Risk Anal. 10(1):85-91.

Swartout, J. 1996. Subchronic-to-chronic uncertainty factor for the reference dose. Abstract F2.03. Society for Risk Analysis Annual Meeting, New Orleans, LA.

Swartout, J.C., P.S. Price, M.L. Dourson, H.L. Carlson-Lynch, and R.E. Keenan. 1998. A probabilistic framework for the reference dose (probabilistic RfD). Risk Anal. 18(3):271-282.

ten Berge, W.F., A. Zwart, and L.M. Appleman. 1986. Concentration-time mortality response relationship of irritant and systemically acting vapours and gases. J. Hazard. Mater. 13(3):301-309.

Travis, C.C., and R.K. White. 1988. Interspecific scaling of toxicity data. Risk Anal. 8(1):119-125.

Van Ryzin, J. 1980. Quantitative risk assessment. J. Occup. Med. 22(5):321-326.

West, J.B., A.R. Elliot, H.J. Guy, and G.K. Prisk. 1997. Pulmonary function in space. JAMA 277(24):1957-1961.

Wilson, J.D. 1997. So carcinogens have thresholds: How do we decide what exposure levels should be considered safe? Risk Anal. 17(1):1-3.

Wiltse, J.A., and V.L. Dellarco. 2000. U.S. Environmental Protection Agency's revised guidelines for carcinogen risk assessment: evaluating a postulated mode of carcinogenic action in guiding dose-response extrapolation. Mutat. Res. 464(1):105-115.

Zeise, L., R. Wilson, and E.A. Crouch. 1987. Dose-response relationships for carcinogens: A review. Environ. Health Perspect. 73:259-306.

Zheng, Q., D.W. Gaylor, and W.K. Lutz. 1995. A carcinogenesis model describing mutational events at the nucleotide level. Proc. Biometrics Sec. Am. Statist. Assoc. 349-354.

5

Ranking Spacecraft Contaminants For Risk Assessment

RESPONSIBILITY for setting priorities and selecting chemicals for risk assessment rests with risk managers at the National Aeronautics and Space Administration (NASA), not with those doing human risk assessments for particular exposures. Setting priorities for risk assessment is a separate function from conducting risk assessments, which in turn is a separate function from making risk management decisions.

There are insufficient resources to conduct formal risk assessments on all potential water contaminants, so priorities must be established among them. The conclusion that this is necessary is based on several assumptions. First, the number of identifiable chemical species, and hence the potential number of hazardous substances in the space station, is large. Second, the number of potential substitutes for these chemicals is larger. Third, there could be several adverse outcomes for any chemical or mixture of chemicals. Finally, in the conduct of a formal risk assessment, the basis for choosing between chemicals and their substitutes, is a substantial exercise, particularly if it requires collecting new experimental data or new information on ambient exposure.

There are many approaches to setting priorities for choosing candidates for formal risk assessment. The approaches range in complexity.

At one extreme, priorities are set by expert judgment. Although expert judgment could result in acceptable choices, the information, assumptions, and logic involved are not necessarily specified. At the other extreme, priorities can be generated by complex and well-specified formulas. Substantial information and resources may be required. Each system has benefits and limitations, and a successful system for ranking candidates for risk assessment will combine approaches in establishing a strategy for establishing priority among chemical candidates that should be subjected to quantitative risk assessment.

APPROACHES TO RISK PRIORITIZATION

We describe three methods for setting priorities below. These examples are meant to describe methods of increasing formality for setting priorities.

Ad Hoc Approach

Candidate chemicals for risk assessment would be proposed as the candidates become of interest to NASA staff. As there would be more candidates than resources available to conduct risk assessments, there would be backlog of chemicals to which new nominees would be added. Periodically, candidate chemicals would be chosen for actual risk assessment. The decision makers would make subjective, qualitative, and, presumably, wise decisions. The parameters or the data elements upon which candidates would be chosen would not be specified, nor would chemical candidates competing for a formal risk assessment be weighed against each other in a quantitative sense. However, in this scenario the decision maker is none the less making difficult and complex choices.

Ad Hoc Approach with Factors Specified

A further step toward formality in setting priorities would be for NASA to specify the parameters it would consider. They might include evidence of exposure, magnitude of routine and accidental exposures, short- and long-term effects, ability to monitor and control exposure, and the need to have the substance on board. An example of a priority system in which the elements are specified but their weights and interrelationships are not specified comes from the National Research Council (NRC) report *Carcinogens and Anticarcinogens in the Human Diet* (NRC 1996). Nine criteria, including "extent of occurrence and use patterns; known human carcinogenicity, but no animal data" are listed. These are based on criteria used by the International Agency for Research on Cancer (IARC 1984) for selecting agents for carcinogenicity testing.

Formal System with Parameters, Weights, and Interrelationships Specified

In a formal system, priorities would be based on a specified set of parameters, a formula would be developed that combines scores for the various parameters, and the relationship and weighting of the parameters would be specified. The formula could be as simple as the sum of scores of various parameters or a more complex formula in which parameters are given unequal weights and their relationship would be other than additive. For example, two parameters could be the possibility for accidental exposure and the relative exposure that represents an immediate danger to life. These two parameters could be weighted equally, or danger to life could be given greater weight. Priority scores might be the mathematical sum of adding scores for the two parameters, weighting them unequally, and then adding them again, or a product derived by multiplying rather than adding. The parameters, weights, and relationships are set by the NASA authority responsible for setting priorities.

An example of a formula-based index for ranking carcinogens is the Permitted Exposure/Rodent Potency (PERP) (Gold et al. 1994, 1997). PERP is the result of division of the permissible lifetime exposure established by the Occupational Safety and Health Administration (OSHA) of the U.S. Department of Labor (in milligrams per kilogram per day) by the lifetime dose that induces tumors in 50% of animals. The PERP was proposed as a method for ranking potentially hazardous exposures rather than a detailed assessment of risk. Similar indices could be developed to facilitate ranking exposures for risk assessment in other areas by using permissible exposure, such as in water, and measures of toxicity from experimental studies.

For example, a list of organic contaminants found in water during a 60-day chamber study is provided in Chapter 2 (Table 2-12). Dividing the value for each contaminant found by a corresponding acceptable level would provide a ratio, similar to a PERP, that could then be as part of a ranking system. There are several possible parameters:

- ***Likelihood of routine exposure:*** Potential for routine exposure is based on a thoughtful examination of the spacecraft environment, including what chemicals are carried onboard; which chemicals are predicted to be produced by routine operations or as the result of short-term experiments; and what exposures occurred on past flights, including frequency, amount of exposure, and severity of consequences. Exposure can be expressed as a multiple of an existing appropriate U.S. Environmental Protection Agency standard, as on OSHA standard, or as a Threshold Limit Value (TLV) established by the American Conference of Governmental Industrial Hygienists.
- ***Medical intelligence:*** Periodically, information will become available from ground-based experience, such as toxicologic testing, or from flight-based experience, such as occurrence of symptoms. This information should be reviewed and used in setting priorities for risk assessment.
- ***Likelihood of unusual exposure:*** Unusual exposures can result from accidents and other unwanted episodes. Many of these exposures are predictable, such as those that would result from fires and leaks.
- ***Severity of toxicity:*** Knowledge of the severity of the toxicity, including considerations of reversibility and ability to perform during and after exposure, is critical.
- ***Design requirements:*** The physical design of the spacecraft should be a major consideration in the selection of chemical candidates for risk assessment. A primary concern is the effectiveness of systems in producing or limiting exposure in the spacecraft to specified levels. Clearly, a decision of what constitutes

acceptable exposure must be considered in the design phase of mechanical and other systems. It could be that the timing of the design phase will in part force selection of chemical species for formal risk assessment. Once the design is completed, and later when the system is built, the system's capacity for controlling and eliminating exposures should determine the choice and use of candidate chemicals during flight.

- ***Special spaceflight considerations:*** The special circumstances and consequences of spaceflight, such as microgravity and its effects on calcium metabolism, should affect choice of candidate chemicals.
- ***Spaceflight experience:*** This includes onboard and ground-based experience.

FLEXIBILITY OF RISK PRIORITIZATION

Flexibility in ranking is encouraged to increase the effectiveness of risk assessment. Flexibility should be manifest in several ways. New information might include changes in the parameters, changes in information about specific agents, and the addition of new substances for consideration. One might foresee a system that produces a series of priority rankings based on changes in parameters considered, their weighting, or their relationships. Priority rankings could then be compared in a way akin to sensitivity analysis in mathematical risk assessment.

SUMMARY

We have described a range of approaches for choosing candidates for risk assessment. One choice is subjective, and the elements considered are not specified by the expert decision makers. In a second approach, the parameters to be considered are specified but their weighting and interrelationships are not. The third approach is more formulaic and involves specifying and quantifying the elements that are considered in the decision as well as the weighting of the elements in the decision making. The subcommittee recommends the use of a combination of these approaches to rank chemicals for risk assessment.

REFERENCES

Gold, L.S., G.B. Garfinkel, and T.H. Slone. 1994. Setting priorities among possible carcinogenic hazards in the workplace. Pp. 91-103 in Chemical Risk Assessment and Occupational Health: Current Applications, Limitations, and Future Prospects, C. M. Smith, D. C. Christiani, and K. T. Kelsey, eds. Westport, CT: Auburn House.

Gold, L.S., T.H. Slone, N.B. Manley, G.B. Garfinkel, L. Rohrbach, and B.N. Ames. 1997. Carcinogenic potency database. Pp. 1-606 in Handbook of Carcinogenic Potency and Genotoxicity Databases, L.S. Gold and E. Zeiger, eds. Boca Raton, FL: CRC Press.

IARC (International Agency for Research on Cancer). 1984. Chemicals and Exposures to Complex Mixtures Recommended for Evaluation in IARC Monographs and Chemicals and Complex Mixtures Recommended for Long Term Carcinogenicity Testing. IARC Intern. Tech. Rep. No. 84/002. Lyon, France: International Agency for Research on Cancer.

NRC (National Research Council). 1996. Carcinogens and Anticarcinogens in the Human Diet. Washington, DC: National Academy Press.

Appendix A

Water Reclamation Systems on *Mir* And the International Space Station

THE descriptions of the *Mir* water systems in this appendix reflect the systems and technology that existed and served as the basis for U.S. experiments during the U.S.–Russian *Mir* Phase-1A program and the technology that the Russians will transfer to the *Mir* segment of the International Space Station.

MIR WATER-RECOVERY SYSTEMS

The *Mir* system for water reclamation and management consists of three isolated loops: one for recovering water from urine, one for recovering potable water from the humidity condensate in cabin air, and one for recovering water from hygiene wastewater. The hygiene loop is not currently in operation and no wash water is recovered. All loops receive supplemental water supplied from Earth, which is transferred manually to each as needed to compensate for losses (Pierre et al. 1999). The urine and wash water recovery loops are located in the Kvant-2 module, and the humidity condensate purification processor is located in the *Mir* core module. In the first loop, water recovered from urine is used in the water electrolysis system to generate oxygen; this water theoretically can be processed further to produce potable water if needed. In the second loop, humidity condensate is used to produce potable water, and it supplies about 80% of the crew's drinking water. The other 20% comes from a water resupply, which contains water produced by fuel cells aboard the space shuttle or water delivered on the Russian water tanks, called Rodnik tanks (210 liter (L) capacity), by the Russian *Progress* resupply vehicle. Having the make-up water supply available has allowed metabolic water balance to be maintained at all times. Each of the Russian systems currently operating in *Mir* is described below.

Humidity Condensate Processing

Most of the potable water consumed aboard *Mir* comes from recycled humidity condensate. This system operated successfully aboard the *Salyut* stations beginning in 1975, and an upgraded version operates on *Mir*. In the *Mir* system, atmospheric humidity condensate is collected and processed into potable water by a condensate water processor located in the core module. Gas and impurities are separated from the liquid (condensates), minerals and disinfectants are added, and the resultant potable water is supplied as hot or cold water for the crew.

The condensate processor unit consists of four treatment subunits: a gas-liquid separator, a multifiltration bed, a conditioning-biocide addition unit, and a distribution and pasteurization system (Figure A-1).

The air-condensate mixture is first passed through a 10-micrometer (μm) filter to remove particulates, and then the gas and liquid are separated by static metal plates that have hydrophilic capillary pores along the wall of the separator. The separated liquid is drawn through the pores by a negative pressure diaphragm pump, and the air is vented to the cabin. Condensate is collected in 180 milliliter (mL) aliquots and pumped to the multifiltration bed.

The multifiltration bed contains ion exchange resins, activated charcoal, and a propriety room-temperature catalyst that remove inorganic and organic contaminants by cationic exchange and oxidation. The catalyst allows removal of low-molecular-weight alcohols, such as ethanol and methanol. An on-line conductivity sensor located down-

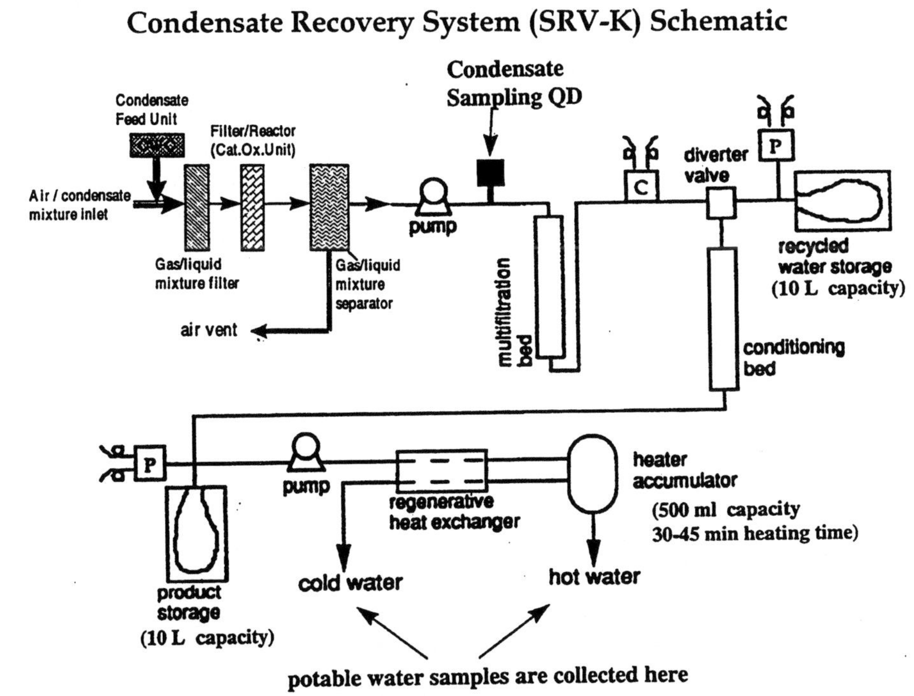

FIGURE A-1 The *Mir* humidity condensate water-reclamation system, which is planned for the early phases of the International Space Station. On the ISS, this system will be located in the service module of the Russian segment. Source: Modified from Pierre et al. (1996).

stream of the multifiltration bed is used to determine whether the water is of acceptable quality (<150 microsiemens per centimeter [μS/cm]). A series of valves directs acceptable water to the conditioning bed, and unacceptable water is sent to a storage tank for recycling. Water that is accepted by the on-line sensor is processed through the conditioning bed, which adds magnesium, calcium, and other minerals to enhance palatability. Biocidal silver (0.05-0.20 milligrams per liter [mg/L]) also is added here for microbial control.

Conditioned water is heated (pasteurized) to 85°C by a heat pump (regenerative heat exchanger) and stored in a heated accumulator. Hot water is available to the crew directly from the accumulator; cold water is provided by rerouting the hot water through the regenerative heat exchanger (Samsonov 1996a).

Filter-Reactor Catalytic Oxidation

The system for water recovery from humidity condensate has been upgraded for the International Space Station (ISS) and includes an assembly that will remove organic contaminants by catalytic oxidation in an air-liquid flow at ambient temperature and pressure upstream of the multifiltration bed. This should at least double the life of the multifiltration beds (Samsonov et al. 1997). The composition of the catalyst and the process is proprietary. In fact, this "filter reactor," an ambient-temperature catalytic reactor, has been in operation aboard *Mir* since January 1998.

In the service module design for the ISS, the system will have a condensate feed unit that will facilitate transfer of condensates collected from ISS modules (and stored in contingency water container bags) to the Russian condensate processor for water recovery.

Urine Processing

A Russian system for reclaiming water from urine has been in operation aboard *Mir* since January 1990. This system was designed primarily for regenerating cabin oxygen through electrolysis. The urine-processing system consists of three parts: urine preparation (collection, preservation-pretreatment, storage), atmospheric distillation, and distillate treatment and purification. These subsystems are described briefly below.

The *Mir* urine preparation subsystem consists of the urinal, a urine-pretreatment assembly, and a blower. Urine is distilled at atmospheric pressure in a distillation unit that consists of an evaporator, condenser, heater, and brine tank. The distillate post-treatment and purification unit is identical to the humidity condensate processor. The urine is treated with sulfuric acid and a liquid solution of a commercially available oxidizer (similar to Oxone, described below), and the treated urine undergoes membrane distillation at atmospheric pressure and at relatively low temperatures. An electric heater raises the temperature to 50-52°C, and water is evaporated with a stream of air that blows at 100 L/min through the evaporator, which consists of a stack of hydrophilic capillary-porous polymeric membranes. The water vapor is condensed in a heat exchanger, cooled by an onboard coolant, and the condensed liquid is processed (post-treated), as described for condensate, except for the water conditioning unit. Urine has never been used to provide potable water on *Mir* missions.

In the life support module (LSM) of the ISS, an upgraded Russian system for processing hygiene water and urine is targeted for deployment in 2002. To meet the needs to increase processing capacity and life span of the ISS configuration and conserve energy, the *Mir* system will be updated for the ISS. The distillation step will receive the most re-engineering. Vacuum distillation with a rotary evaporator-condenser and a thermoelectric heat pump will be the principal component of this new system (Samsonov et al. 1996a,b). Apart from water for electrolysis, a subsystem (the water-recovery system–urine subsystem) will be added for conditioning, distributing, and preheating of the water to process it to potable quality if needed (Samsonov et al. 1997).

U.S. Water-Processing System

The U.S. water-reclamation system is a single-loop system to produce potable water from a mixture of urine distillate, humidity condensate, and hygiene (wash) water, consisting of body and hand-washing water. (Laundry was deleted in the transition between the space station *Freedom* and the ISS program; similarly, a shower stall also has been eliminated even though the water-processing system will be designed to process water for personal hygiene.) Particulate filtration, adsorption, ion exchange, catalytic oxidation, and biocide addition are done in various subsystems. The U.S. water-processing system hardware for the ISS has been tested over the past several years in isolated and integrated modes at NASA's Marshall Space Flight Center. This test bed, the Water Recovery Test, has greatly facilitated ground-based evaluation of individual systems, components, and processes of water reclamation. The proposed U.S. integrated water-reclamation system is illustrated in Figure A-2; details of its subsystems are provided below.

Urine Processing

Urine and urinal flush water are collected, treated, and distilled by vapor compression before this mixture enters the water-processing sys-

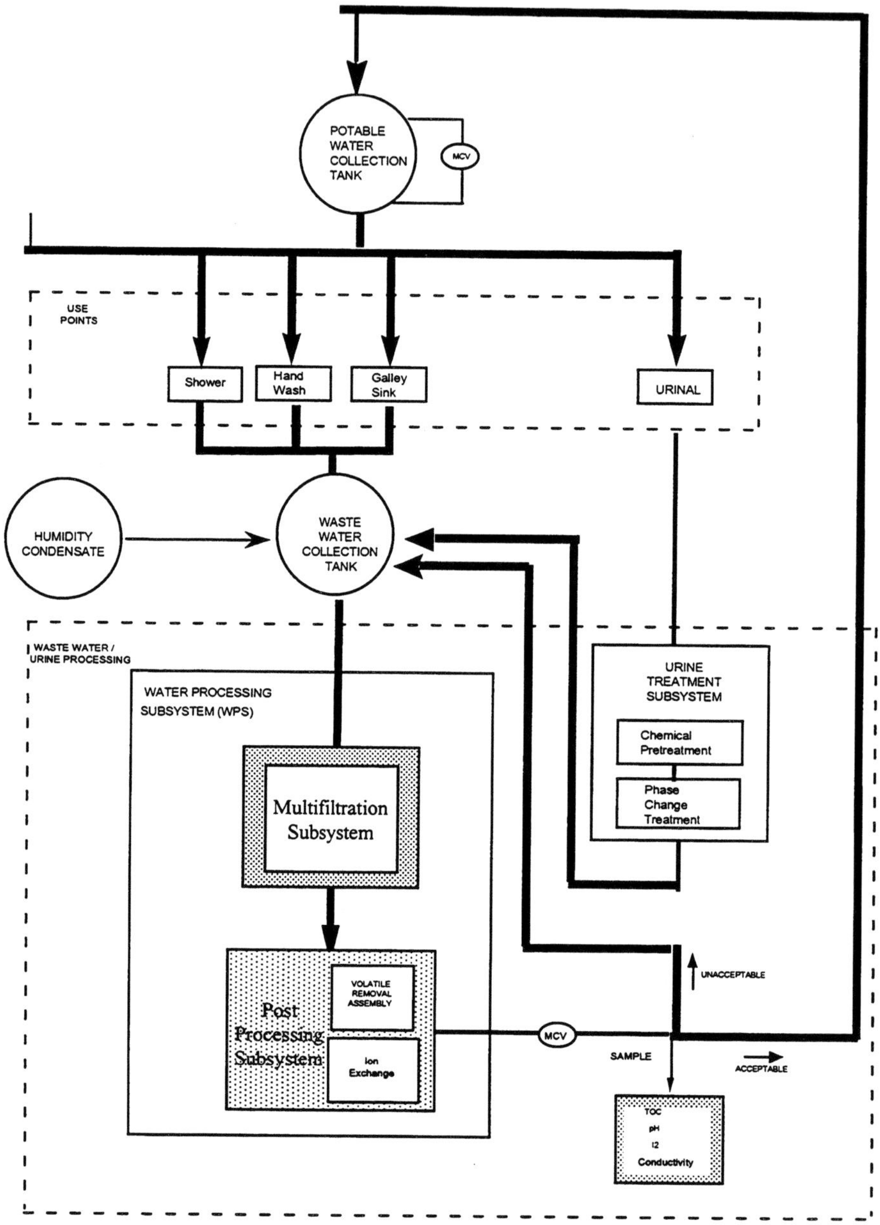

FIGURE A-2 U.S. water-reclamation system for the ISS. MCV, microbial check valve; TOC, total organic carbon. Source: Verostko et al. (1997).

tem (Figure A-2). The collected urine and flush water first are stabilized to control microbes and fix ammoniated species in solution by the addition of Oxone (a proprietary potassium monopersulfate compound) and sulfuric acid. The urine and flush water are sent on for vapor compression distillation (VCD). VCD was chosen over other technologies for ISS after ground-based tests revealed that the VCD apparatus weighed less, required less power and less filtration, and produced better quality water than did other equipment. In VCD treatment, the stabilized urine is evaporated at low pressure to form water vapor, which is sent to a compressor that increases condensation temperature and pressure.

The compressed vapor is directed to the condenser. The latent heat produced by condensation is transferred to provide heat for evaporation. The process of evaporation, compression, and condensation takes place between 32°C and 38°C. The resultant brine is recirculated through a recycle loop to maximize the amount of water that can be extracted. Concentrate from the system is collected in a tank within the VCD unit. If the conductivity of the distillate (measured by a conductivity sensor in the VCD unit) exceeds 120 µS/cm, it is reprocessed; if the conductivity is acceptable, the effluent is sent via a wastewater distribution line to the combined wastewater processor upstream of a particulate filter.

Combined Urine and Wastewater Processing

The major portion of the U.S. water reclamation system processes product water from the VCD unit (urine distillate), humidity condensate (which includes condensate from human metabolism and from off-gassing), and wash water (from the shower, hand wash, and oral hygiene). These influents are stored in a stainless steel tank and delivered to the processing system under pressure (8 pounds per square inch gauge) from a feed pump. The back pressure created on the waste distribution line by the storage unit facilitates the removal of gas, which is released into the cabin. Wastewater is passed through a 0.5-µm filter to remove particulate contaminants, and then continues to the multifiltration subsystem and on to the volatile removal assembly, as shown in Figure A-3.

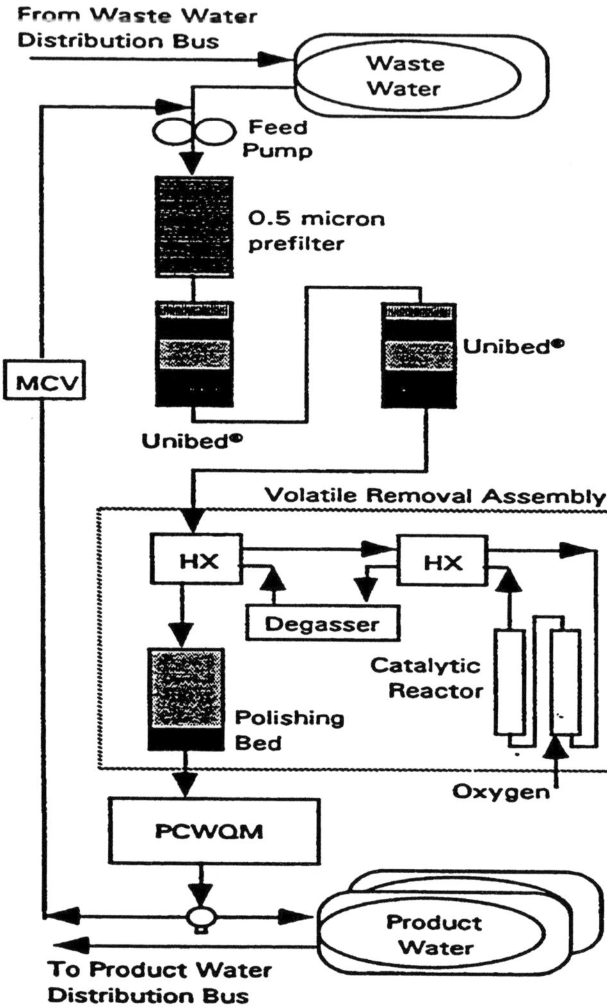

FIGURE A-3 ISS Integrated U.S. potable-water processing system. HX, heat exchanger; MCV, microbial check valve; PCWQM, process-control water quality monitor. Source: Holder et al. (1995).

Multifiltration Subsystem

The multifiltration subsystem consists of two Unibed filtration units plumbed in series. Each contains various adsorbents and ion exchange resins chosen for their ability to remove specific organic and inorganic contaminants expected to be present in the wastewater. The media in these units include iodinated anion exchange resin, strong and weak cation and anion exchange resins, activated carbon from coconut shell and coal, and polymeric adsorbent. Conductivity sensors are located at the multifiltration inlet, between the two Unibeds, and at the multifiltration outlet. These sensors are used to monitor Unibed performance so the units can be replaced as necessary.

The Unibeds cannot remove low-molecular-weight or polar organic compounds, such as ethanol or urea. Thus, the effluent is sent to the volatile removal assembly (VRA) for further treatment.

Volatile Removal Assembly

The VRA consists of regenerative heat exchangers, an oxygen sparger, a catalytic reactor, a gas-liquid separator, and a "polishing" ion exchange bed (Figure A-3). The VRA oxidizes organic compounds from the multifiltration effluent to carbon dioxide. This oxidation takes place at moderate temperature. The feed water is heated twice, once as it flows through a regenerative heat exchanger and again by an immersion-type heater, and it reaches about 130°C before it enters the catalytic reactor. The oxidizing conditions and the moderate temperature help to maintain microbial contamination at less than 100 colony-forming units per 100 mL of water. After catalytic oxidation, the feed water passes back through the regenerative heat exchanger for heat reclamation before passing through a polishing ion exchange resin, which removes organic acids and other ionic contaminants. A microbial check valve adds 2-4 mg/L of iodine (and 1 mg/L of iodide) to the effluent water as a residual disinfectant. The membrane-based gas-liquid phase separator helps remove excess gas (oxygen) and other gaseous oxidation by-products from the partially cooled effluent from the reactor.

Requirements, Proposed Limits, and Monitoring

Tables A-1, A-2, and A-3 present NASA's potable-water quality requirements for the ISS (Table A-1), compare NASA and Russian pro-

TABLE A-1 NASA Potable-Water Quality Requirements (Maximum Contaminant Levels) for the International Space Station

Parameter (units)	Levels
Physical Parameters	
Total solids (mg/L)	100
Color, true (Pt-Co units)	15
Taste (TTN)	3
Odor (TON)	3
Particulates (maximum size: μm)	40
pH	6.0-8.5
Turbidity (NTU)	1
Dissolved gas (free at 37°C)	ND[a]
Free gases (at STP)	ND[a]
Inorganic Constituents, mg/L[b,c]	
Ammonia	0.5
Arsenic	0.01
Barium	1.0
Cadmium	0.005
Calcium	30
Chlorine (total, includes chloride)	200
Chromium	0.05
Copper	1.0
Iodine (total, includes organic iodine and iodide)	15
Iron	0.3
Lead	0.05
Magnesium	50
Manganese	0.05
Mercury	0.002
Nickel	0.05
Nitrate (NO_3-N)	10
Potassium	340
Selenium	0.01
Silver	0.05
Sulfate	250
Sulfide	0.05
Zinc	5.0
Bactericide, mg/L	
Residual iodine (minimum)	1.0
Residual iodine (maximum)	4.0
Aesthetics, mg/L	
Cations	30
Anions	30
CO_2	15
Microbial	
Total count, CFU/100 mL	
(bacteria[e]/fungi[f])	100[d]
Total coliform[g]	ND[d]

TABLE A-1 *(Continued)*

Parameter (units)	Levels
Virus	ND[h]
Radioactive Constituents (pCi/L)[I]	
Organic Parameters (μg/L)[b]	
Total acids	500
Cyanide	200
Halogenated hydrocarbons	10
Total phenols	1
Total alcohols	500
Total organic carbon (TOC)	500
Uncharacterized TOC[j]	100
Organic Constituents (mg/L)[b,c]	

[a]ND, no detectable gas using volumetric gas versus fluid measurement system. Excludes CO_2 used for aesthetic purposes.

[b]Each parameter or constituent must be considered individually and independently of others.

[c]In the event a quality parameter not listed in this table is projected, or found, to be present in the reclaimed water, the water quality manager at Johnson Space Center will be contacted for a determination of the MCL for that parameter.

[d]Membrane filtration method.

[e]Incubation time: 48 hr; temp.: 30 °C; medium: R2A.

[f]Incubation time: 48 hr; temp.: 30 °C; Medium: DG-18.

[g]ND, not detectable. Incubation time: 24 hr; Temp.: 35°C; Medium: M-Endo.

[h]Tissue culture assay.

[i]The MCLs for radioactive constituents in potable and personal hygiene water are to conform to Nuclear Regulatory Commission regulations (10 CFR 20). The MCLs are listed in the *Federal Register*, Vol 51, No. 6, 1986, Appendix B, as Table 2 (Reference Level Concentrations), Column 2 (Water). Control, containment, and monitoring of radioactive constituents used are the responsibility of the user. Before the introduction of any radioactive constituents approval is to be obtained from the Radiation Constraints Panel, which will approve or disapprove proposed monitoring and decontamination procedures on a case-by-case basis.

[j]Uncharacterized TOC equals TOC minus the sum of analyzed organic constituents expressed in equivalent TOC.

CFU, colony forming units; NTU, nephelometric (turbidity) units; Pt-Co, platinum-cobalt scale; STP, standard temperature and pressure; TTN, threshold taste number; TON, threshold odor.

Source: Adapted from SSP 50005 Rev B (1995). NASA Space Station Program. This information appears in a table listed as Figure 7.2.2.3.2-1.

posed contaminant limits for potable water aboard the ISS (Table A-2), and outline the schedule requirements for monitoring water quality (Table A-3).

TABLE A-2 Comparison of Proposed Limits for Potable Water Aboard the ISS

Parameter	Maximum Contaminant Levels	
	NASA	Russian Space Agency
pH[a]	5.5-9.0	5.5-9.0
Color[a]	15 Pt-Co units	20 degrees
Taste[b]	3 TTN	2 points
Odor[b]	3 TON	2 points
Total dissolved solids[c]	100, 1000 mg/L	100, 1000 mg/L
Turbidity[b]	1 NTU	1.5 mg/L
Total gas	5% volume at ATM, 20°C	5% volume at ATM, 20°C
Ammonia (NH_4-N)	1.5 mg/L	2 mg/L
Arsenic	0.01 mg/L	0.01 mg/L
Barium	1 mg/L	1 mg/L
Cadmium	0.005 mg/L	0.005 mg/L
Calcium	30 mg/L	100 mg/L
Chlorine, total (includes Cl^-)	200 mg/L	250 mg/L
Chromium	0.1 mg/L	0.1 mg/L
Copper	1 mg/L	1 mg/L
Fluorine	1.5 mg/L	1.5 mg/L
Iodine, total (includes I^-)	15 mg/L	0.05 mg/L
Iodine, residual[d]	1.0-4.0 mg/L	NA
Iron	0.3 mg/L	0.3 mg/L
Lead	0.05 mg/L	0.05 mg/L
Magnesium	50 mg/L	50 mg/L
Manganese	0.05 mg/L	0.05 mg/L
Mercury	0.002 mg/L	0.002 mg/L
Nickel	0.1 mg/L	0.1 mg/L
Nitrate (NO_3-N)	10 mg/L	10 mg/L
Selenium	0.01 mg/L	0.01 mg/L
Silver	0.5 mg/L	0.5 mg/L
Sulfate	250 mg/L	250 mg/L
Zinc	5 mg/L	5 mg/L
Total hardness (Ca, Mg)	7 meq/L	7 meq/L
Cyanide	200 μg/L	200 μg/L
Total phenols	1 μg/L	1 μg/L
Ethylene glycol	12 mg/L	12 mg/L
Total organic carbon (TOC)	500 μg/L	20,000 μg/L[e]
Uncharacterized TOC	100 μg/L	No limit
Oxygen consumption - COD	No limit	100 mg
Total bacteria[b]	100 CFU/100 mL	10,000 CFU/100 mL
Coliform bacteria	<1 CFU/100 mL	<1 CFU/100 mL
Virus	<1 PFU/100 mL	<1 PFU/100 mL

Agreements reached at the Joint Working Group meeting (Feb. 9-13, 1998) for the shuttle-*Mir* and ISS water supply and water quality.

[a]pH range applies only before iodination.

TABLE A-2 *(Continued)*

[b]Different values for U.S. and Russian-supplied water. To be further reviewed.

[c]The 100 mg/L limit applies to the water before mineralization; thereafter, total dissolved solids may not exceed 1000 mg/L.

[d]Range is applicable if iodine is the biocidal agent. (Silver is used as the biocide in the Russian program.)

[e]This limit does not include the mineral counter-ion, formate.

ATM, atmosphere; CFU, colony-forming units; COD, coefficient oxygen delivery; meq/L, milliequivalent per liter; NA, not applicable; NTU, nephelometric (turbidity) units; PFU, plaque-forming units; Pt-Co, platinum-cobalt scale; TTN, threshold taste number; TON, threshold odor.

TABLE A-3 Schedule Requirements for Water Quality Monitoring

	On-line[a]	Off-line[b]	
Measurement	Potable	Potable	Hygiene
Physical			
Total solids	—	—	—
Color	—	+	+
Conductivity	x	x	x
Taste and odor	—	+	+
Particulates	—	+	+
pH	x	x	x
Temperature	x		
Turbidity	—	+	+
Dissolved gas	—	+	
Free gas	—	+	
Inorganic Compounds			
Ammmonia	—	+	+
Iodine	x	x	x
Specific contributors[c]	—	+	+
Aesthetic			
Specific contributors[d]	—	+	+
Microbial			
Total count (bacteria, fungi)	—	x	x
Total coliform	—	x	x
Virus	—	—	—
Microbe ID[e]	—	x	x
Radionuclide[f]			
Organics			
TOC	x[g]	x	x
Specific organics[c]	—	+	+

x, monitoring required; —, monitoring not required; +, monitoring requirement will be waived if verification testing and analysis indicate that the quality measure limit will be met reliably.

[a]Process-stream samples will be analyzed to provide real-time or near-real-time results for process control and for presumptive water quality assessment. Requirements for on-line monitoring of additional parameters will be established if verification testing and analysis indicates that such monitoring is required for process control or water quality assessment.

[b]In addition to the on-line and off-line analyses, grab samples from the water systems will be obtained for later ground analysis.

[c]Specification of organic and inorganic elements and compounds to be monitored will be based on the potential for those elements and compounds to be present in the product water and on their toxicity. If a parameter not listed in this table is projected or found to be present in the reclaimed water, the water quality manager at the Johnson Space Center will be contacted to determine monitoring requirements.

[d]Selection will be based on determination of critical aesthetic parameters.

[e]Does not include identification of viruses.

TABLE A-3 *(Continued)*

[f]The ability to monitor radionuclides during flight will be provided as part of the experiment or procedure that involves their use.

[g]Analytical procedure could provide an indirect equivalent of classical TOC.

Source: SSP 50005 Rev B, August 1995, Figure 7.2.7.3.2.1-1; International Space Station Flight Crew Integration Standard NASA STD 3000T.

REFERENCES

Holder, D.W., D.L. Carter, and C.F. Hutchens. 1995. Phase III Integrated Water Recovery Testing at MSFC: International Space Station Configuration Test Results and Lessons Learned. SAE Technical Paper Series no. 951586, 25th International Conference on Environmental Systems, San Diego, CA, July.

Pierre, L.M., J.R. Schultz, S.M. Johnson, R.L. Sauer, Y.E. Sinyak, V.M. Skuratov, and N.N. Protasov. 1996. Collection and Chemical Analysis of Reclaimed Water and Condensate from Mir Space Station. SAE Technical Paper Series no. 961569, 26th International Conference on Environmental Systems, Monterey, CA, July 8-11.

Pierre, L.M., J.R. Schultz, R.L. Sauer, Y.E. Sinyak, V.M. Skuratov, N.N. Pratasov, and L.S. Bobe. 1999. Chemical Analysis of Potable Water and Humidity Condensate: Phase One Final Results and Lessons Learned. SAE Technical Paper Series no. 1999-01-2028. 29th International Conference on Environmental Systems, Denver, CO, July 12-15.

Samsonov, N.M., L.S. Bobe, V.M. Novikov, N.S. Farafonov, G.K. Abramov, B.Y. Pinsky, E.I. Grigorov, E.N. Zaitsev, N.N. Protasov, V.V. Komolov, A.I. Grigoriev, and Y.E. Sinyak. 1996a. Water Supply Based on Water Reclamation from Humidity Condensate and Urine on a Space Station. SAE Technical Paper Series no. 961408, 26th International Conference on Environmental Systems, Monterey, CA, July 8-11.

Samsonov, N.M., L.S. Bobe, V.M. Novikov, B.Y. Pinsky, N.V. Rykhlov, V.A. Soloukhin, N.S. Farafonov, N.N. Protasov, V.V. Komolov, V.G. Rifert, and V.V. Rakov. 1996b. Problems of Developing Systems for Water Reclamation from Urine for Perspective Space Stations. SAE Technical Paper Series no. 961409, 26th International Conference on Environmental Systems, Monterey, CA, July 8-11.

Samsonov, N.M., L.S. Bobe, V.M. Novikov, N.S. Farafonov, G.K. Abramov, B.Y. Pinsky, E.I. Grigorov, E.N. Zaitsev, N.N. Protasov, V.V. Komolov, A.I. Grigoriev, and Y.E. Sinyak. 1997. Updated Systems for Water Recovery from Humidity Condensate and Urine for the International Space Station. SAE Technical Paper Series no. 972559, 27th International Conference on Environmental Systems, Lake Tahoe, NV, July 14-17.

SSP (Space Station Program). 1995. International Space Station Flight Crew Integration Standard NASA 3000T Document (Aug.). water50005 Rev B. Johnson Space Center, Houston, TX.

Verostko, C.E., K. Pickering, F. Smith, N. Packham, J. Lewis, G. Stonesifer, D. Staat, and M. Rosenbaum. 1997. Performance of the Water Recovery System During Phase II of the Lunar-Mars Life Support Test Project. SAE Technical Paper Series no. 972417, 27th International Conference on Environmental Systems, Lake Tahoe, NV, July 14-17.

Appendix B

Benchmark Dose Estimation

BELOW are examples of how to derive benchmark doses (BMDs) using data on 1,4-dichlorobenzene, botulinum, vinyl chloride, and aflatoxin.

DERIVING BENCHMARK DOSES FOR 1,4-DICHLOROBENZENE

In the examples provided below, the subcommittee used two 13-week (wk) toxicity studies and one 2-year (yr) study by the National Toxicology Program (NTP 1987) as the basis for deriving BMDs. Two 13-wk studies were done because the first study did not demonstrate a no-observed-adverse-effect level (NOAEL). In the first study, rats were administered 1,4-dichlorobenzene by gavage at doses of 300-1500 mg/kg (milligrams per kilogram body weight) per day (d), 5 d/wk. Because histologic changes of the kidney were observed in male rats in all dose groups, a second study was done at lower doses of 38-600 mg/kg/d. In the 2-yr study, 1,4-dichlorobenzene was administered by gavage 5 d/wk at 150 and 300 mg/kg/d for male rats and at 300 and 600 mg/kg/d for female rats.

The results of these studies are used below to illustrate the way to calculate BMDs for nonquantal response data, quantal data that are highly variable, and for carcinogenic effects (Table B-1). The tabulated benchmark dose (BMD_p), which is a lower 95% confidence limit (CL)

TABLE B-1 Toxicity Data and Benchmark Doses for 1,4-Dichlorobenzene

Hematology: Decreased Hematocrit in Male F-344/N Rats (13-wk study)

Dose (mg/kg)	0	300	600	900	1200	1500
Avg. hematocrit (%)	50.1 ± 0.8	47.8 ± 0.9	47.3 ± 0.7	47.2 ± 0.4	47.6 ± 0.8	42.5 ± 0.5 (±SE)
Sample size	9	9	9	9	5	2

Model-Free (Experimental Dose)			Normal Distribution, 3 SD Adverse Effect		
NOAEL	LOAEL	LOAEL/NOAEL	BD_{01}	BD_{10}	BD_{10}/BD_{01}
None	300	NA	112	334	3

Hematology: Decreased Hemoglobin Concentration in Male F-344/N Rats (13-wk study)

Dose (mg/kg)	0	300	600	900	1200	1500
Hemoglobin (g/dL)	17.6 ± 0.2	16.4 ± 0.2	16.5 ± 0.2	16.5 ± 0.2	16.6 ± 0.3	15.3 ± 0.2 (±SE)
Sample size	9	9	9	9	5	2

Model-Free (Experimental Dose)			Normal Distribution, 3 SD Adverse Effect		
NOAEL	LOAEL	LOAEL/NOAEL	BD_{01}	BD_{10}	BD_{10}/BD_{01}
None	300	NA	75	203	3

Hematology: Decreased RBC Count in Male F-344/N Rats (13-wk study)

Dose (mg/kg)	0	300	600	900	1200	1500
RBC (1,000,000/mm^3)	10.0 ± 0.12	9.5 ± 0.11	9.5 ± 0.12	9.7 ± 0.09	9.8 ± 0.17	8.8 ± 0.16 (±SE)
Sample size	9	9	9	9	5	2

Model-Free (Experimental Dose)			Normal Distribution, 3 SD Adverse Effect		
NOAEL	LOAEL	LOAEL/NOAEL	BD_{01}	BD_{10}	BD_{10}/BD_{01}
None	300	NA	81	230	3

Body Weight: Decreased Weight in Male F-344/N Rats (13-wk study)

Dose (mg/kg)	0	300	600	900	1200	1500
Avg. weight gain (g)	187 ± 5	163 ± 8	142 ± 5	137 ± 6	114 ± 9	91 ± 22 (±SE)
Sample size	9	10	10	9	5	2

Model-Free (Experimental Dose)

NOAEL	LOAEL	LOAEL/NOAEL
None	300	NA

Normal Distribution, 3 SD Adverse Effect

BD_{01}	BD_{10}	BD_{10}/BD_{01}
87	232	3

Kidney: Renal Tubular Degeneration in Male F-344/N Rats (13-wk study)

Dose (mg/kg)	0	37.5	75	150	300	600
Response	7/10	NR	NR	5/10	3/10	9/10
Sample size	9	9	9	9	5	2

Severity increased at 300 mg/kg and 600 mg/kg; 150 mg/kg is considered a NOAEL

Model-Free (Experimental Dose)

NOAEL	LOAEL	LOAEL/NOAEL
150	300	2

Probit Model (ad hoc 95% LCL)

BD_{01}	BD_{10}	BD_{10}/BD_{01}
72	190	3

Liver: Hepatocellular Adenoma or Carcinoma in Male $B6C3F_1$ Mice (2-yr study)

Dose (mg/kg)	0	300	600
Response	17/50	22/49	40/50

Model-Free (Experimental Dose)

NOAEL	LOAEL	LOAEL/NOAEL
8	77	10

Probit Model (ad hoc 95% LCL)

BD_{01}	BD_{10}	BD_{10}/BD_{01}
9	43	5

LCL, lower confidence limit; LML, linearized multistage model; LOAEL, lowest-observed-adverse-effect level; NOAEL, no-observed-adverse-effect level; NA, not applicable; NR, not reported; RBC, red blood cell; SD, standard deviation.

estimate of a dose corresponding to an excess risk of p, has not been adjusted for a discontinuous exposure regimen (dosing was done 5 d/wk). For the calculation of spacecraft water exposure guidelines (SWEGs), a duration conversion would have to be made along with other necessary conversions and adjustments.

Nonquantal Response Data

In the first 13-wk toxicity study, NTP reported hematotoxicity in male rats at all doses. Specifically, hematocrit, hemoglobin concentration, and red blood cell (RBC) count were decreased at all doses relative to the vehicle control. None of these hematologic changes was seen consistently in female rats. (Hematotoxicity was not mentioned in regard to the second 13-wk study.) Decreased body weight gain was observed at all doses in male rats, in parallel to hematotoxicity. No NOAEL was determined for decreased hematocrit, decreased hemoglobin concentration, decreased RBC count, or decreased body weight gain in male rats, and, thus, the lowest dose tested (300 mg/kg) was considered the lowest-observed-adverse-effect level (LOAEL). The data on these four parameters are used for illustrating the calculation of BMD for nonquantal response data.

The data on each of the hematologic parameters and on body weight were modeled using the procedure of Kodell and West (1993) to calculate BMD. This method assumes a normal distribution for the observed end point, with a quadratic dose-response function for the average response. For the calculations here, a response was considered adversely low if it fell more than three standard deviations below the theoretical average response for the vehicle control. Because of substantially reduced survival at the two highest doses (1200 and 1500 mg/kg, 50% and 20% survival, respectively), data from those groups were not used for model fitting. Figures B-1, B-2, and B-3 display each hematology data set plotted against the mean dose-response model estimated by maximum likelihood. Figure B-4 gives similar information for the body weight gain data.

BMD_{01} and BMD_{10} for decreased hemoglobin concentration, decreased RBC count, and decreased body weight gain are all in close agreement. The BMD_{01} is 75, 81, and 87 mg/kg, respectively; the BMD_{10} is 203, 230, and 232 mg/kg, respectively. Each BMD_{10} is reason-

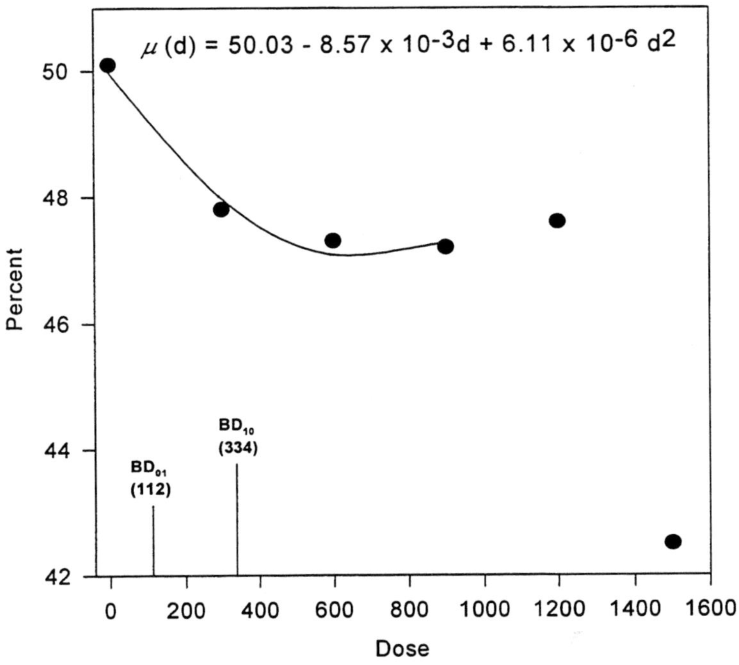

Figure B-1: Hematocrit data modeled using the procedure of Kodell and West (1993).

ably close to the corresponding LOAEL (300 mg/kg) for the toxic effect (Table B-1). For decreased hematocrit, the resulting BMD_{01} (112 mg/kg) and BMD_{10} (334 mg/kg) are higher than the corresponding BMD_p for other hematotoxic effects and for decreased body weight gain. To determine the effect of including the two highest dose groups in the calculation of BMD_p for hematocrit, the BMD_p was recalculated for all 6 dose groups. The resulting BMD_{01} was 250 mg/kg and the BMD_{10} was 746 mg/kg—both of which seem unrealistically high. The modeling procedure does not provide a goodness-of-fit test, but it seems likely that the model does not fit well when all dose groups are included. Certainly, the response at the highest dose appears to have too much influence on the predictions at low doses. It seems prudent,

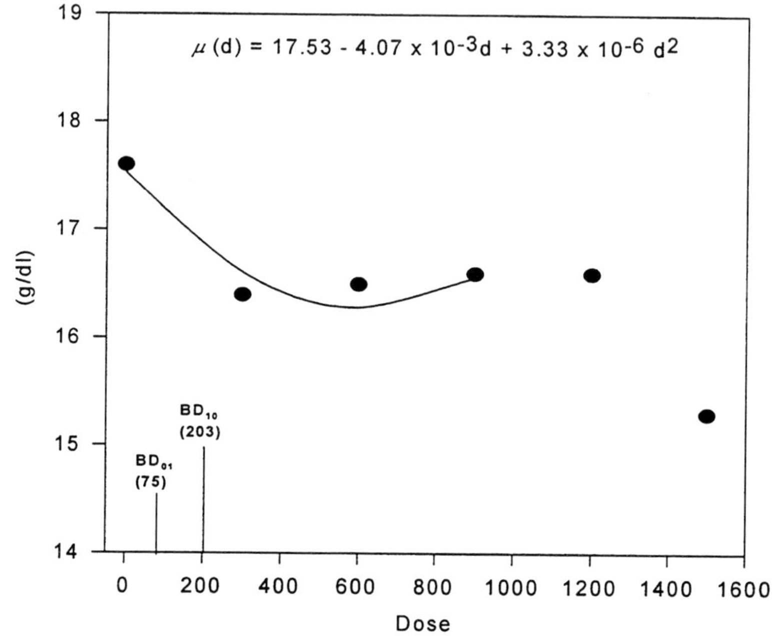

Figure B-2: Hemoglobin concentration data modeled using the procedure of Kodell and West (1993).

in this case, to confine BMD calculations to the four dose groups with good survival by eliminating the two highest dose groups. As Figures B-1, B-2, and B-3 indicate, the quadratic models for mean hematology responses fitted to the four lowest dose groups are not monotone with respect to dose. One could fit monotonic dose-response models, but such models could be less steep and would not fit the data as well. The result, as with the inclusion of the highest doses, could be overestimation of BMD_p.

Highly Variable Quantal Data

The second 13-wk study of 1,4-dichlorobenzene identified a NOAEL of 150 mg/kg in male rats, based on renal tubular degeneration, which

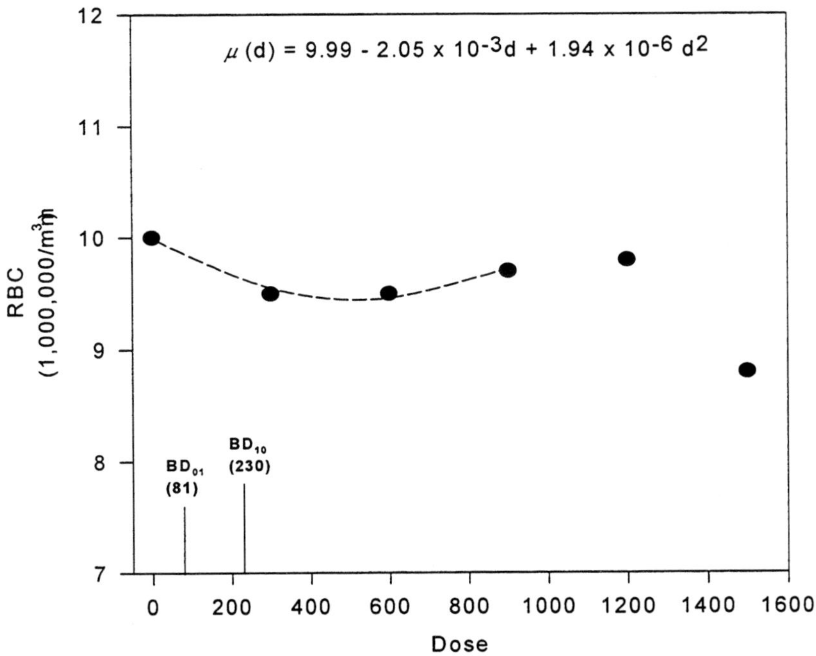

Figure B-3: Red blood cell count modeled using the procedure of Kodell and West (1993).

occurred at doses of 300 mg/kg and higher. (This also confirmed the results of the first study.) The data exhibited anerratic dose-response relationship: Degeneration was mild in 7 of 10 rats in the control group; mild to moderate in 5 of 10 rats in the 150-mg/kg group; moderate in 3 of 10 rats in the 300-mg/kg group; and moderate in 9 of 10 in the 600-mg/kg group (responses at 37.5 mg/kg and 75 mg/kg were not reported).

Based on *increased severity* of renal tubular degeneration at 300 mg/kg, this dose was identified as a LOAEL. There was no increase in severity at 150 mg/kg, and because of the numerically lower incidence compared with the vehicle control, this was considered a NOAEL. A probit log-dose model was fitted to the incidence data, and a goodness-of-fit test indicated a marginally acceptable fit (p = .07)—the fitted

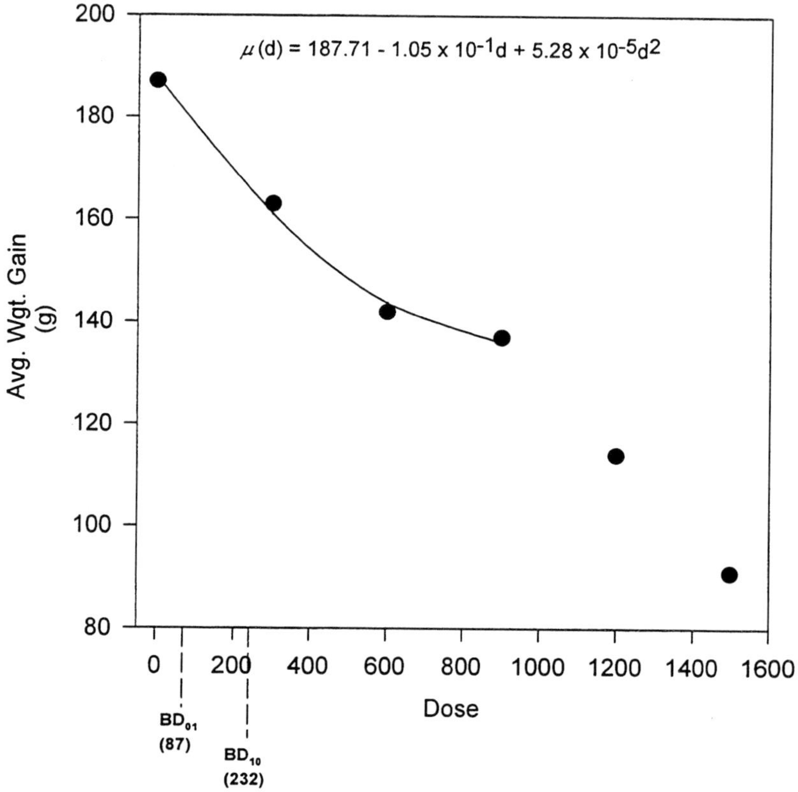

Figure B-4: Body weight gain modeled using the procedure of Kodell and West (1993).

model did not differ statistically from the data at the 5% significance level. However, because of the high variability in the data that describe the dose-response relationship as shown in Figure B-5, convergence of the maximum likelihood estimation procedure could not be achieved for calculating lower CLs on doses corresponding to the specified excess risks of 1% and 10% (BMD_{01} and BMD_{10}). Therefore, the ad hoc method (outlined in Chapter 4 in the section on Experimental Variation) was used to derive factors by which to reduce central estimates of dose to calculate lower CLs. The resulting BMD_{01} is 72 mg/kg (449/6.2), which is about one-half of the NOAEL (150 mg/kg),

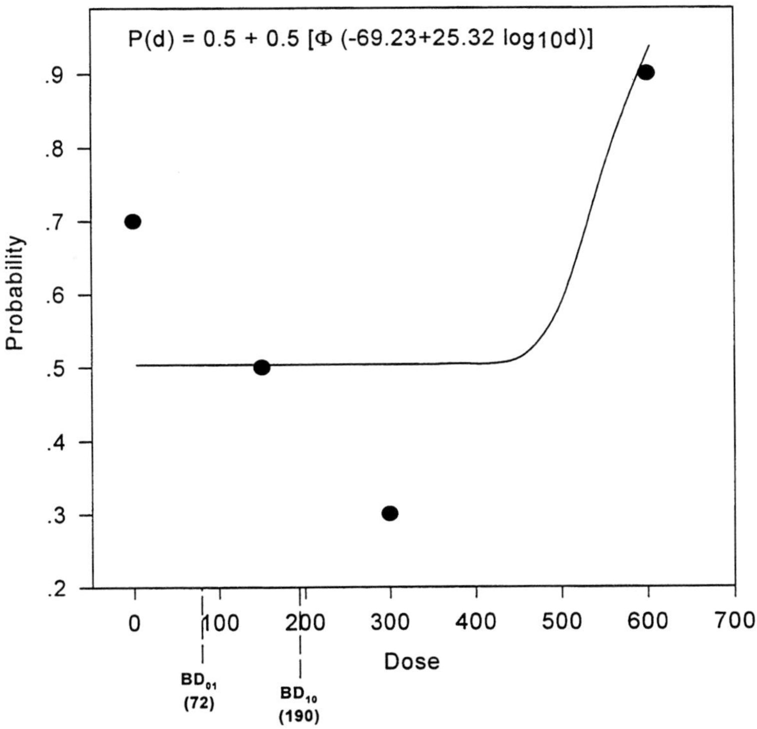

Figure B-5: Renal tubular degeneration data modeled using a probit log-dose model.

and the BMD_{10} is 190 mg/kg (494/2.6), which is about two-thirds of the LOAEL (300 mg/kg) (Table B-1). The dose-response relationship for these data are perhaps too erratic to be used as a basis for SWEGs (Figure B-5). However, the BMD_{01} and BMD_{10} are reasonably close to the corresponding values calculated for the three hematology end points and decreased body weight gain (Table B-1).

Carcinogenic Effects

NTP found clear evidence of carcinogenicity in the 2-yr study. In male rats, there was an increased incidence of renal tubular cell adenocarcinomas. In male and female mice, there were increased incidences of hepatocellular adenomas and carcinomas. Because the U.S. Environmental Protection Agency (EPA) used hepatocellular adenomas and carcinomas in male mice to calculate a carcinogenic potency factor for 1,4-dichlorobenzene, the data are used here to illustrate the calculation of BMDs for carcinogenic effects.

Two separate calculations were done, one using the (linearized) multistage model and the other using the probit log-dose model (Table B-1, Figures B-6 and B-7). The BMD_{01} values are in close agreement (multistage: 8 mg/kg; probit: 9 mg/kg). The BMD_{10} from the multistage model (77 mg/kg) is about 2-fold higher than the BMD_{10} from the probit model (43 mg/kg). The BMD_{01}s for carcinogenic effects are almost a factor of 10 lower than the lowest BMD_{01} for noncancer effects. For the calculation of SWEGs for water aboard spacecraft, these BMD_{01}s would need to be reduced further by an interspecies uncertainty factor and possibly by a factor to reflect the severity and irreversibility of cancer. If, for example, two factors of 10 were applied, then the resulting SWEG would be equivalent to a value that the National Aeronautics and Space Administration (NASA) would calculate as a spacecraft maximum allowable concentration (SMAC) for cancer, in that it would correspond to a linearly extrapolated excess risk of 10^{-4}.

It should be noted that the cancer BMD_{01} of 8 mg/kg produced by the multistage model can easily be related back to EPA's cancer potency factor of 2×10^{-2}, assuming linearity of the dose-response curve at low doses. To define the potency factor, one must adjust 8 mg/kg to account for the discontinuous exposure regimen used in the study by multiplying by 5/7 (dosing was done 5 d/wk), and then divide by an interspecies surface area adjustment factor of approximately 13. This gives 0.44 mg/kg for a BMD_{01} that is adjusted for duration and species. Because $0.01 = 0.44 \times$ potency, assuming linearity, then potency $\cong 2 \times 10^{-2}$.

Discussion

It is interesting to note the ratios of LOAELs to NOAELs and BMD_{10}s to BMD_{01}s in Table B-1. All ratios lie between 2 and 10. If BMD_{01}s are to be used instead of NOAELs and BMD_{10}s are to be used instead of

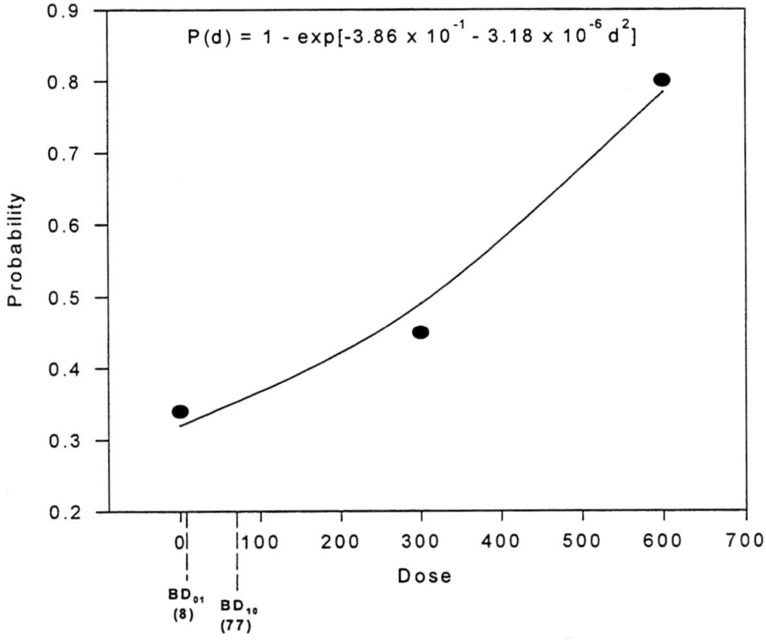

Figure B-6: Multistage model of hepatocellular adenoma or carcinoma incidence.

LOAELs in the calculation of SWEGs, then it would be better to use a BMD_{01} whenever possible, rather than starting with a BMD_{10} and, perhaps, having to reduce it by as much as 10-fold.

Another ratio of interest for comparison between BMD_{01}s and BMD_{10}s is the ratio of the model-based central estimate of dose to its corresponding lower 95% CL. For renal tubular degeneration, because of the ad hoc way the lower CL were calculated, it is easy to see that the ratio for BMD_{01} is at least 2-fold higher than the ratio for BMD_{10} (449/72 = 6.2 versus 494/190 = 2.6). For hepatocellular tumors, the ratio for BMD_{01} is 3-fold higher than the ratio for BMD_{10} (e.g., multistage model: 56/8 = 7.0 versus 181/77 = 2.3). Likewise, for decreased hematocrit, the ratio for BMD_{01} is higher than the ratio for BMD_{10} (203/112 = 1.8 versus 398/334 = 1.2). However, for decreased hemoglobin concentration, decreased RBC count, and decreased body weight gain, the ratio for BMD_{01} is smaller than the corresponding ratio

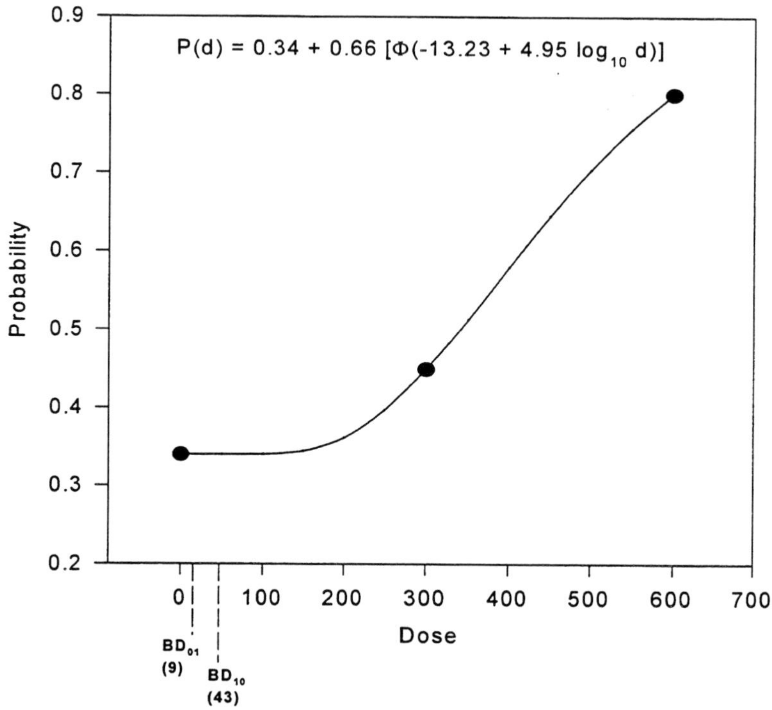

Figure B-7: Probit log-dose model of hepatocellular adenoma or carcinoma incidence

for BMD_{10} (111/75 = 1.5 versus 332/203 = 1.6, 126/81 = 1.6 versus 491/230 = 2.1, and 134/87 = 1.5 versus 367/232 = 1.6, respectively). Thus, it is not possible to predict that the ratio of the model-based central estimate of dose to its corresponding lower 95% CL will always be larger for one value of *p* than the other.

DERIVING BMDs FOR BOTULINUM TOXIN, VINYL CHLORIDE, AND AFLATOXIN

Data on one toxic end point from botulinum toxin (lethality), vinyl chloride (liver tumor), and aflatoxin (liver tumor) are used to further illustrate the calculation of model-based BMD_{01}s and BMD_{10}s, and to make comparisons to the corresponding model-free NOAELs and LOAELs (Table B-2). Each data set illustrates a particular type of dose-response relationship, and each highlights interesting characteristics of BMD_p calculations.

A probit log-dose model was used to fit the botulinum toxin data in Table B-2 (Food Safety Council 1980), assuming a background response level of zero. Using all data points resulted in a significant lack of fit of the model ($p < .05$). Thus, the highest doses were eliminated one at a time until a satisfactory fit was achieved ($p = .76$) (Figure B-8). This resulted in the elimination of the three highest dose groups. The elimination of high dose groups is a generally accepted practice because the focus is on getting the best possible estimates in the low-dose region of interest. Whether all doses below 27 pg (picograms) were kept or discarded made little difference in the fit of the model and its predictions; thus, these doses were kept for the analysis. The resulting 1% and 10% benchmark doses are given in Table B-2 along with the NOAEL and LOAEL, for purposes of comparison. The LOAEL was established at 30 pg, because the response at this dose level was statistically different ($p < .05$) from the zero response at all lower doses, including the NOAEL of 27 pg.

For the botulinum toxin data, the dose-response relationship is very steep, and very thresholdlike (Figure B-8). For this reason, the BMD_{01} and BMD_{10} are close, and they mimic their counterparts, the NOAEL and LOAEL, fairly well.

For the vinyl chloride data in Table B-2 (Food Safety Council 1980), all dose groups except the 1 ppm (part per million) group were from the same experiment. The 1 ppm group was part of a later experiment conducted under the same conditions as the first. A probit log-dose model was used to fit these data, assuming a background response level of zero. An excellent fit was obtained ($p = .79$). The resulting 1% and 10% BMDs are given in Table B-2, along with corresponding NOAELs and LOAELs. Although 250 ppm is the lowest dose that is statistically different from control ($p < .05$), the nonzero response at 50 ppm makes it difficult to decide which of the two doses (50 ppm or 250 ppm) should be established as the LOAEL. The data and maximum-likelihood fitted model are shown in Figure B-9.

The BMD_{01} and BMD_{10} are far apart for the vinyl chloride data, differing by a factor of 285. This is because of the one-hit nonthreshold shape of the dose-response curve. Excess risk is predicted all the way

TABLE B-2 Toxicity Data and Benchmark Doses for Botulinum Toxin, Vinyl Chloride, and Aflatoxin

Botulinum Toxin: Lethality

Dose (pg)	1	5	10	15	20	24	27	30
Response	0/10	0/10	0/30	0/30	0/30	0/30	0/30	4/30
Dose (pg)	34	37	40	45	50	55	60	65
Response	11/30	10/30	16/30	26/30	26/30	17/30	22/30	20/30

Model-Free (Experimental Dose)			Probit Log-Dose Model (95% LCL)		
NOAEL	LOAEL	LOAEL/NOAEL	BD_{01}	BD_{10}	BD_{10}/BD_{01}
27	30	1.1	22	28	1.3

Vinyl Chloride: Liver Tumor

Dose (ppm)	0	1	50	250	500	2500	6000
Response	0/58	0/118	1/60	4/59	7/60	13/60	13/59

Model-Free (Experimental Dose)			Probit Log-Dose Model (95% LCL)		
NOAEL	LOAEL	LOAEL/NOAEL	BD_{01}	BD_{10}	BD_{10}/BD_{01}
1 or 50	50 or 250	50 or 5	0.9	257	285

Aflatoxin B_1: Liver Tumor

Dose (ppb)	0	1	5	15	50	100
Response	0/18	2/22	1/22	4/21	20/25	28/28

Model-Free (Experimental Dose)			Probit Log-Dose Model (95% LCL)		
NOAEL	LOAEL	LOAEL/NOAEL	BD_{01}	BD_{10}	BD_{10}/BD_{01}
150	300	2	72	190	3

LCL, lower confidence limit; LOAEL, lowest-observed-adverse effect level; NOAEL, no-observed-adverse-effect level. Source: Food Safety Council (1980).

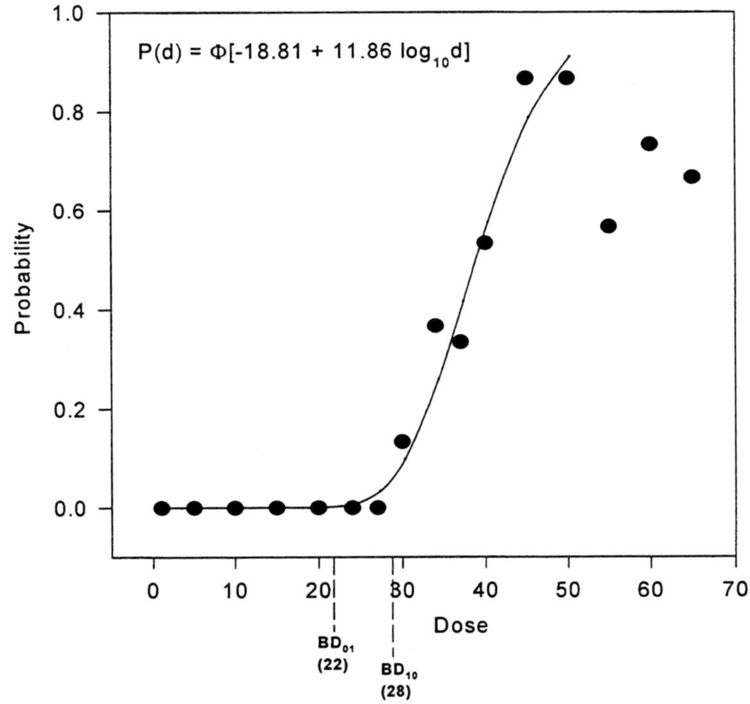

Figure B-8: Probit log-dose model of botulinum toxin lethality data.

down to zero dose (Figure B-9), so that the BMD_{01} is low. For these data, the BMD_{01} and BMD_{10} do not agree well with their counterparts, the NOAEL and LOAEL. If the NOAEL is taken to be 1 ppb (parts per billion), then there is good agreement with the BMD_{01} of 0.9 ppb, but the LOAEL of 50 ppb does not agree well with the BMD_{10} of 257 ppb. On the other hand, if the NOAEL is taken to be 50 ppb, then it does not agree well with the BMD_{01} of 0.9 ppb, but the LOAEL of 250 ppb agrees well with the BMD_{10} of 285 ppb. This case illustrates one advantage of using BMD_p instead of NOAEL, that is, no subjective judgment is needed to determine precisely which experimental dose is a NOAEL. Rather, fitting a model to the dose-response data enables estimation of any dose as a BMD_p.

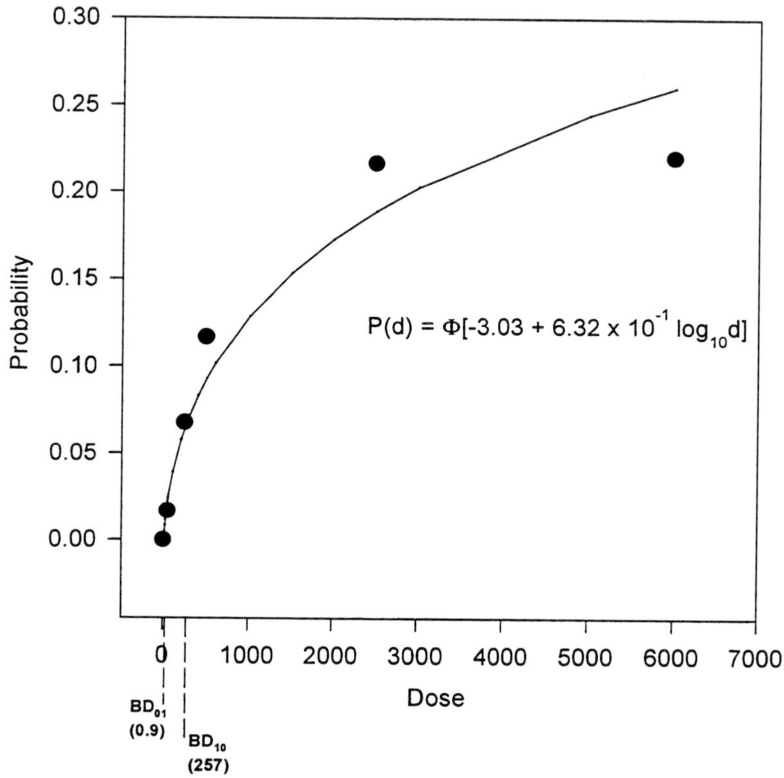

Figure B-9: Probit log-dose model of liver tumor data on vinyl chloride

A probit log-dose model was fitted to the aflatoxin B_1 data in Table B-2 (Food Safety Council 1980), without any assumption about background response. Thus, a nonzero background was allowed. A good fit was obtained (p = .30) (Figure B-10). The resulting BMD_p estimates are given in Table B-2, along with the NOAEL (5 ppb) and LOAEL (15 ppb). The LOAEL was 15 ppb because that was the lowest dose for which the response was statistically different from the zero response at zero dose ($p < .05$).

For the aflatoxin data, the BMD_{01} and BMD_{10} are roughly half the

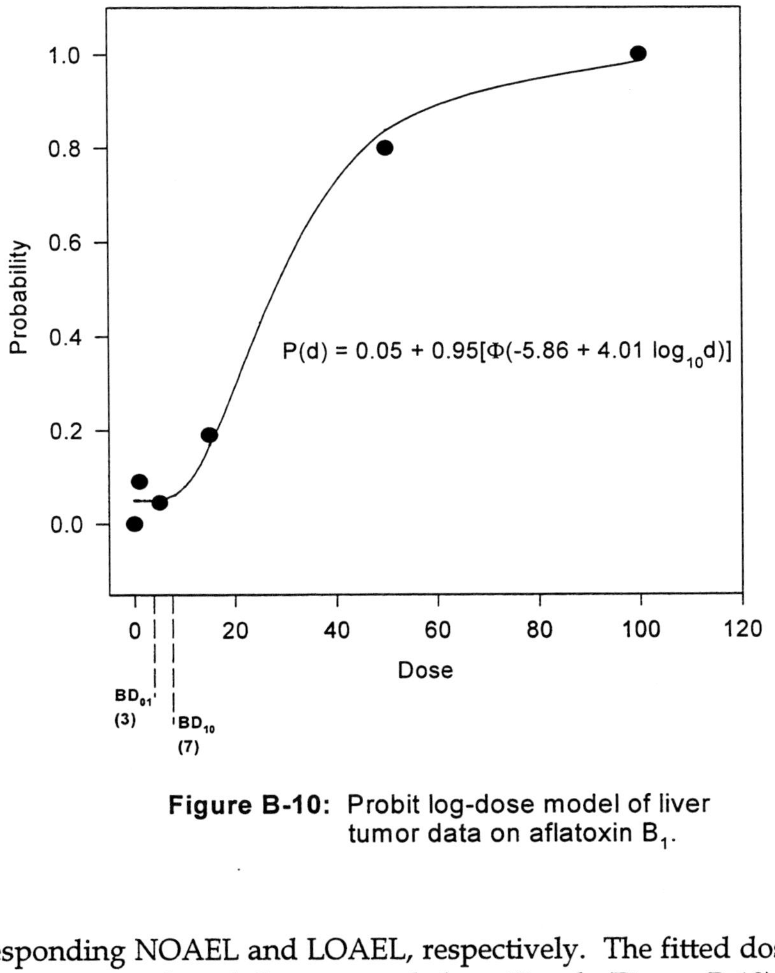

Figure B-10: Probit log-dose model of liver tumor data on aflatoxin B_1.

corresponding NOAEL and LOAEL, respectively. The fitted dose-response is sigmoid, and flattens out below 15 ppb (Figure B-10). Because of the inversion of the observed responses at 1 ppb (9%) and 5 ppb (4.5%), the model predicts a nonzero background rate of approximately 5%, even though the observed background rate is zero. Although the data are quite variable at low doses, they are consistent with a nonzero background rate, and the model does provide a good fit to the data. It appears reasonable that the BMD_{01} and BMD_{10} are 2-fold lower than the corresponding NOAEL and LOAEL, because they reflect the variability of the dose-response relationship at low doses (the response at 1 ppb is numerically higher than is the response at 5 ppb).

REFERENCES

Food Safety Council. 1980. Proposed System for Food Safety Assessment. Washington, DC: Food Safety Council.

Kodell, R.L., and R.W. West. 1993. Upper confidence limits on excess risk for quantitative responses. Risk Anal. 13:177-182.

NTP (National Toxicology Program). 1987. Toxicology and Carcinogenesis Studies of 1,4-dichlorobenzene (CAS No. 106-46-7) in F344/N Rats and $B6C3F_1$ Mice (Gavage Studies). Technical Report No. 319. Research Triangle Park, NC: U.S. Department of Health and Human Services.

Appendix C

Biographical Information on the Subcommittee on Spacecraft Water Exposure Guidelines

DONALD E. GARDNER *(Chair)* is president of Inhalation Toxicology Associates, Inc., a consulting firm in inhalation toxicology. He received his Ph.D. in environmental health at the University of Cincinnati. His research interests include environmental and occupational toxicology, immunotoxicology, pulmonary toxicology, and host defense mechanisms.

JOSEPH V. BRADY is professor of neuroscience and director of the Behavioral Biology Research Center at the Johns Hopkins University School of Medicine. He received his Ph.D. in behavioral biology from the University of Chicago. His research interests include experimental analysis of behavior, behavioral physiology, behavioral pharmacology/toxicology, and space flight performance studies.

GARY P. CARLSON is professor of toxicology and associate head in the School of Health Sciences at Purdue University. He received his Ph.D. in pharmacology from the University of Chicago. His research interests are primarily related to the relationship between the bioactivation of chemicals and their toxic actions. Of particular interest are solvents and alcohols.

ELAINE M. FAUSTMAN is professor in the Department of Environmental Health and director of the Institute for Risk Analysis and Risk Communication at the University of Washington. She received her Ph.D. in pharmacology and toxicology from Michigan State University. Dr. Faustman's research interests include mechanistic investigations of the reproductive and developmental toxicity of metals and she has developed quantitative risk assessment methods for noncancer end points.

CHARLES E. FEIGLEY is professor of environmental health sciences at the University of South Carolina School of Public Health. He received his Ph.D. in environmental sciences and engineering from the University of North Carolina. Dr. Feigley's research is primarily in the areas of exposure assessment and occupation hygiene engineering with an emphasis on developing innovative exposure assessment and control methods.

MARY ESTHER GAULDEN is adjunct professor of radiology at the University of Texas Southwestern Medical Center. She received her Ph.D. in biology from the University of Virginia and did pre-doctoral research at the National Institutes of Health and postdoctoral research at Oak Ridge National Laboratory. Dr. Gaulden's research interests include the effects of low doses of radiation, especially on chromosomes and on man, and the genotoxicity and carcinogenesis of chemicals.

WILLIAM E. HALPERIN is a senior scientist at the National Institute for Occupational Safety and Health in Cincinnati. He received his M.D., M.P.H., and Dr.P.H. from Harvard University. His research interests include occupational epidemiology, public health surveillance, and occupational medicine.

RALPH L. KODELL is director of biometry and risk assessment at the Food and Drug Administration's National Center for Toxicological Research. He received his Ph.D. from Texas A&M University. His research interests include developing and applying mathematical models and statistical techniques in the design and analysis of laboratory experiments to predict human risk.

ROBERT SNYDER is professor and chair of the department of pharmacology and toxicology at Rutgers University College of Pharmacy, and is associate director of the Environmental and Occupational Health Sciences Institute. He received his Ph.D. in biochemistry from the State University of New York at Syracuse. His research interests include solvent toxicology, chemically induced bone marrow depression, liver toxicity, and chemical carcinogenesis.

BERNARD M. WAGNER is president of Wagner Associates, Inc., a consulting firm in toxicology and pathology. He is also emeritus research professor of pathology at New York University Medical Center and serves as a consultant to national and foreign government agencies, academia, and industry. He received his M.D. from Hahnemann Medical College. His research interests include toxicology, diseases of the connective tissue, and comparative pathology.

GAROLD S. YOST is professor of pharmacology and toxicology at the University of Utah. He received his Ph.D. in organic chemistry from Colorado State University. Dr. Yost's research interests include the study of drug metabolism and mechanisms of pneumotoxic chemicals.